WATER FOR ENERGY NEXUS, FOR DEVELOPMENT OF AFRICA

Water for Energy Nexus, for Development of Africa

Swati

KALPAZ

ISBN : 978-9351-2818-87
Price : ₹ 700

First Published, 2016

Published by

Kalpaz Publications
C-30, Satyawati Nagar,
Delhi-110052
Ph. : 9212142040
E-mail: kalpaz@hotmail.com

Printed at : G. Print Process, Delhi

1. Citation for the front cover map "Product of the Transboundary Freshwater Dispute Database, College of Earth, Ocean, and Atmospheric Sciences, Oregon State University. Additional information about the TFDD can be found at <http://www.transboundarywaters.orst.edu>"
2 Citation for the back cover map : The citation information: Onyeji, Ijeoma. 2014. "Chapter 3 - Harnessing and Integrating Africa's Renewable Energy Resources A2 - Jones, Lawrence E." In Renewable Energy Integration, 27–38. Boston: Academic Press. http://www.sciencedirect.com/science/article/pii/B978012407910600003X.
3. World Bank, Energy Department map, 2007

Cataloging in Publication Data—DK

Courtesy: D.K. Agencies (P) Ltd. <docinfo@dkagencies.com>

Swati, author.

Water for energy nexus, for development of africa / Swati.

pages cm
Includes bibliographical references.

1. Energy security—Africa. 2. Water security—Africa. 3. Water-supply—Africa. 4. Energy policy—Africa. I. Title.

HD9502.A35S78 2016 DDC 333.79096 23

Contents

List of Figures

Acronyms

AU: African Union

SADC: Southern African Development Community

ECOWAS: Economic Community of West African States

WAPP: West African Power Pool

SAPP: Southern African Power Pool

UNDP: United Nations Development Programme

GEF: Global Environment Facility

SSA: Sub Saharan Africa

DRC: Democratic Republic of Congo

NAMREP: Namibian Renewable Energy Programme

SAWEP: South African Wind Energy Programme

PRSP: Poverty Reduction Strategy Papers

ODA: Official Development Assistance

NEPAD: New Partnership for Africa's Development

GDP: Gross Domestic Product

GWP: Global Water Partnership

KTOE: Kilo Tons Oil Equivalent

GWh: Gigawatt Hour

PV: Photovoltaics

KW: Kilowatt

REASWA: Renewable Energy Association of Swaziland

RENAMO: Resistência Nacional Moçambicana (The Mozambican National Resistance)

ZRA: Zambia Revenue Authority

OKACOM: Okavango River Basin Water Commission

BSAP: Biodiversity Strategy, Action Plan and National Report

CHM : Clearing House Mechanism

UNFCC: United Nations Framework Convention on Climate Change

WAP: W-Arly-Pendjari

NCSA: National Capacity Self-Assessment

COP: Conference of the Parties

LDC: Least Developed Countries

POPs: Persistent Organic Pollutants

NAPA: National Adaptation Programme of Action

LADA: Land Degradation Assessment in Dryland Areas

ASCLME: Agulhas and Somali Current Large Marine Ecosystems

ADRAR: Appui a la formulation du projet de l'association

BARREM: Barrier Removal to Malawi Renewable Energy Programme

BCLME: Benguela Current Large Marine Ecosystem

CCA: Common Country Assessment

CALLC: Capacity through Local Level Coordination

COGERAT: Joint Management of the Aîr and Ténéré Reserve

PLECO: Micro-basins in the Departments of Gouré and Mainé-Soroa

SAP: Strategic Action Programme

SWH: Solar Water Heating

ABI: Agulhas Biodiversity Initiative

CBD: Convention of Biological Diversity

Glossary

Nationalisation of Nature: To render distinctively national or governmental character, scope or notoriety of nature or an ecosystem.

Naturalisation of Nation: The act of investing an alien with the rights and privileges of a native in this case Transboundary rights for river basin or water management.

Water Conflicts: These conflicts with respect to this paper are being less traditional, increasingly being driven by internal, local or trans-boundary pressures, or more subtle issues of poverty and stability.

"Internationalised" basins: The basins whose management institution was developed under a single jurisdiction, but which was altered or shattered as that jurisdiction suddenly became divided among two or more nations.

Resilience: The capacity of an individual, community or system to adapt in order to sustain an acceptable level of function, structure, and identity and resume original form after going through major changes such as severe droughts, civil wars, or deforestation as in the case of this report.

Water Security: It is the emerging paradigm that recognises the need for sufficient water for economic and social development to sustain and enhance important ecosystem functions.

Energy Security: It refers to a resilient energy system, which includes following parameters for a thriving economy: 1. Reduced dependence on external supplies, 2. Diversity of energy supply, 3. Functional energy markets, 4. Stable Infrastructure for energy generation, transmission, and distribution, and 5. Environmental sustainability.

Transboundary waters: It refers to sources of freshwater that are shared among multiple user groups, with diverse values and different needs associated with water use.

Hydro political vulnerability: It is the risk of political dispute over shared water systems.

Acknowledgements

This study was performed as an apprenticeship assignment titled "Energy for Water" with UN Habitat Energy and Transport Section; Water, Sanitation, Infrastructure Branch (WSIB), Human Settlements Financing Division (HSFD) under the supervision of Dr Kitio, Energy Advisor based in UN Habitat Nairobi office. Overall supervision was also provided by Daniel Adom Chief Technical Officer Energy.

The author is grateful for the assistance of numerous who assisted in providing information towards making of this report and its review. Particularly, the author wants to thank those who provided guidance from UNDP, GEF, World Bank and UNIDO.

Author expresses gratitude to Mr Kjorven, Deputy Administrator for UNDP and Vice Chair for UN Energy for generous guidance to receive accurate information on the current state of affairs of UNDP/GEF energy projects in Africa. Author also thanks Ms Monique Barbut, GEF Executive Director's office as well as Dr Khandeh Yumkella, UNIDO Director General's office for substantive input for this study.

Author expresses deep gratitude and acknowledgement to Dr John Boright, Executive Director, The National Academies for his support and feedback for successfully completing the report.

Author acknowledges Dr Andrew Hudson, Principal Technical Advisor, International Waters and Cluster Leader, Water Governance Programme on Water, UNDP for providing suggestions towards accuracy of the document; Minoru Takada, Manager, Sustainable Energy Programme, UNDP for providing generous guidance in form of exhaustive references of existing work done by the UN agencies in the area of energy, Alastair Morrison, member of UNDP Water Governance Facility/ Stockholm International Waters Initiative for proof reading the document and providing generous technical guidance for accuracy of this document, Dr Aaron Wolf, Professor of Geography, the Department of Geosciences, Oregon State University and co-director of University Partnership on Transboundary Waters for providing assistance around relevant documents on Transboundary waters, for proof reading and providing technical advice on the document.

We also thank Prof. Wolf for giving us maps for the study particularly the one we have used at the back cover. Author acknowledges Dr James McMahon and Dr Robert Van Buskirk, U.S. Lawrence Berkeley Laboratory for sharing valuable insights regarding their experiences around implementing renewable energy technologies in Africa and providing relevant references towards this end. Author also thanks Dr. Helena Chum, US National Renewable Energy Laboratory, National Bio-energy Center for sharing valuable information and references regarding progress done in bio-energy area around the world and its applicability for Africa.

Author is thankful to all those who provided guidance to develop this study and believes that it will be of value towards Africa's growth and development.

Swati

Abstract

This study is a sincere effort to understand the interdependency of challenges, emerging markets and vulnerability issues faced by Africa, in the area of water and energy sector, that has hindered the economic growth and it's much expected 'green revolution' in the continent. It does so by studying the water conflicts in the region and its implications on electricity generation and economic development. These water conflicts in the continent, the study believes; might be responsible for massive deforestation, exodus of internal migration in search of fertile land, soil erosion and disruption of growth. It concludes that strategic energy investment by the GEF programmes, UNDP and World Bank in the continent will not automatically bring in economic development implying that some additional intervention or efforts are needed in areas such as governance, enabling environment, integrated policies, efficient management, and effective regulatory framework.

Introduction

Experts say that nowhere in the world are challenges greater than in Africa. This study tries to address how the interdependency of these challenges in the area of water and energy has impacted the economic growth in the continent. It further makes an attempt to understand how water conflicts have risen out of its scarcity and impacted the energy production, in turn affecting the resiliency in Africa. **Resiliency** is what the study is looking at – Africa's resiliency towards ecosystem, economy and bureaucracy to address these challenges.

Africa consists of 63 international river basins and many water demands. Studies show that more than 600 lakes in Africa have shrunk dramatically over the past decades, drained by deforestation, pollution and farming. Dams, sewage and industrial pollution have had huge impacts on its fisheries. In past 20 years countries such as Niger have lost more than 80 per cent of their freshwater wetlands[4], Lake Chad once the fourth largest African lake is on the verge of extinction, and future alterations in Chobe River's shape and position could potentially impact Botswana and Namibia's water resources availability[5].

Africa has been through its own share of droughts, civil wars, and energy shortages and they haven't still prioritised the inter-dependency linkage between water resources and energy shortage to its fullest. This kind of ignorance has lead

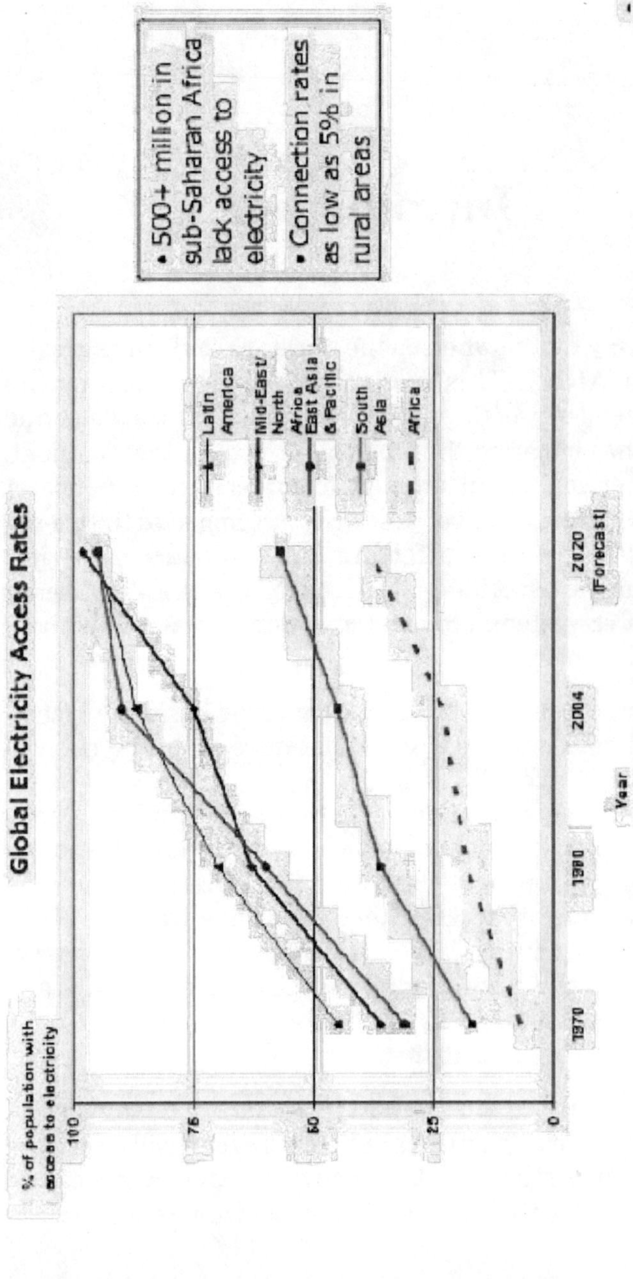

Source: Africa Energy Unit World Bank Presentation by Mac Cosgrove-Davies, Sr. Energy Specialist.

Figure 1. Global Electricity Access Rate,

to heavy prices on electricity supply and failure to use water efficiently leading to depleting resources in Africa. This should not necessarily encourage energy subsidies, which might discourage conservation of water and forests as well as investment in renewable energy options.

It is therefore important for the African states, to integrate energy use into water-planning that could save money, reduce waste, mitigate deforestation and strengthen its economy. Since water is used throughout the energy sector, including in resource extraction, refining and processing, electric power generation, storage, and transport, they should have integrated energy policies that will address water quality due to waste streams, runoff from mining operations, water used from oil and gas extraction, and air emissions affecting downwind watersheds.

African Union/NEPAD should therefore consider facilitating a water policy with embedded energy savings that might result into altered crop patterns resulting in energy savings through harvesting, cultivating and processing or changing the course of irrigation or water allocation patterns for industrial development. The energy implications of this water policy will extend far beyond the energy embedded in water.

The biggest challenge for Africa to understand its variable water availability and energy capacity is having access to accurate information pertaining water consumption, increased demand and availability, policy coordination to make effective decisions regarding water and drought management. This has directly impacted continent's ability to effectively combat climate change – a culprit in this context.

The African Union/NEPAD should consider looking at the importance of collecting water data on the basis of watersheds or drainage basins, and electricity supply and demand usually projected on the basis of census divisions and power grid boundaries, although the two are often incongruent. This is particularly important to understand the needs of water intense energy applications such as extraction, mining, refining, and processing for energy production, generation, and use. Africa should make an effort to come up with its own water-energy research roadmap to address steadily decreasing water resource.

Due to uncontrolled rate of population growth in Africa, the rate of consumption has largely been unchecked and has outpaced available water aquifers, rendering it unable to get accurate data. Water storage is an important aspect of any water infrastructure – for both generation of electricity and supply of freshwater. Africa is greatly lacking in this aspect inspite of a strong river system. The importance of co-producing energy and water by locating power plants next to water treatment facilities to recycle the brackish water for non-potable means hasn't been fully capitalised in the continent.

Africa has not taken stock of an integrated approach of the impact of appropriate water policies and regulations on energy supplies while those of energy policies and regulations on water supplies and demands. It therefore hasn't been able to quantify energy and water resources in order to secure energy security and sustainability. The core of managing energy savings is through a "system context, where significant improvements in energy and water conservation can often be realised through implementation of innovative processes or technologies, co-location of energy and water facilities, or improvements to energy and water infrastructures[6]." This can also assist in maintaining natural resources and ecosystem balance in Africa by using renewable energy option.

In 2007[7], the largest country shares of renewables annual investment were in Germany, China, the United States, Spain, Japan, and India. Germany invested over $14 billion, in mostly wind and solar PV; China invested $12 billion, mostly in small hydropower, solar hot water, and wind power; United States was number three, with over $10 billion investment and Africa has barely invested few million dollars in this sector.

The Critical Issues Related to Energy and Water Security for Africa

In a business-as-usual scenario, as the population increases rapidly, the consumption of water in the electric sector will grow substantially. This will increase the demand for water providing an incentive for technologies that reduce water use, thus dampening the increase in water use. The trends in energy use, water availability, and overuse of water will force Africa to take a fresh look at its development, investment and utilisation means and ways to sustain the increase in population. Most of the 90 per cent of African economies rely heavily on hydropower or traditional biomass as its primary source of energy. That means fluctuations in rainfall[8] and decrease in forest produce will result into a national security issue for some 45 economies.

Inspite of this, only 15 countries are having renewable energy policy and only 50 per cent are having national energy policies. It is forecasted that about 25 countries in Africa will be experiencing extreme severe shortages in energy supply in coming twenty years. Often national policies are developed to promote an area or sector such as oil recovery for domestic energy supply or use of traditional biomass as an alternative energy option for those economies that don't have access to fossil fuels for domestic purpose, causing unintended impacts on the water availability and quality or deforestation and soil erosion respectively. It is understood that there will always be competition for water between different users as is seen so dramatically in Africa and the proven way to reduce these trans-boundary conflicts in this situation is investments in innovative processes and technologies; dams and reservoirs, through integrated infrastructure assessment in a sound energy policy.

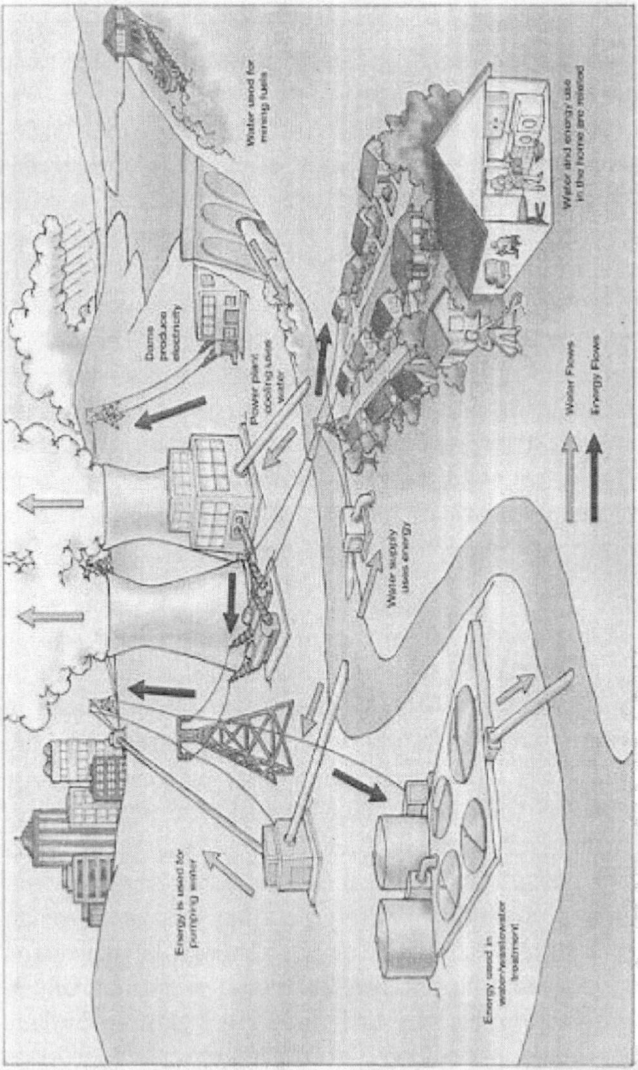

Reference: 2006 US Congress report on Interdependency of Energy and Water.

Figure 2. Examples of Interrelationship between energy and water.

Africa has failed to understand its hydrology and a bigger mistake to politicise it that has lead to disowning the continent as one union. This very well might be the cause of so many trans-boundary conflicts – neighbouring countries not being resilient enough to tackle the magnitude of conflicts with others and with themselves. For example, South African troops moved into Angola to occupy and defend the Ruacana hydropower complex, including the Govè Dam on the Kunene River. The goal was to take possession of and defend the water resources of southwestern Africa and Namibia.

Inspite of terrible losses in trans-boundary water conflicts, value of water has increased leading to increased energy costs, making way to creating incentives for decreasing water intensity for energy production such as renewable energy.

Interaction between water and energy system is critical for Africa – a vulnerable infrastructure system such as leaky water pipes with no water meters, has lead to loss of water and higher bills. Low water levels due to droughts and trans-boundary conflicts have limited the ability of producing power to its full potential in Africa, while increasing its dependence on water pumps and desalination of freshwater aquifers. For example, declining rainfall has led to growing fights between animal herders and farmers with competing needs. In August, 2007 people were forced to flee their homes by fighting in Zounweogo province of Burkina Faso.[9] Dams and water transfers in the South African area of the Incomati river basin reduced freshwater flows and increased salt levels in Mozambique's Incomati estuary leading to disappearance of salt-intolerant plants and animals important to the local livelihood.

Annual renewable freshwater resource levels have depleted rapidly exceeding the precipitation levels across Africa. It has to therefore alter ways it does business throughout the energy sector. For Africa the primary source of energy has been hydropower – whose over exploitation

Energy Element	Connection to Water Quantity	Connection to Water Quality
Energy Extraction and Production		
Oil and Gas Exploration	Water for drilling, completion, and fracturing	Impact on shallow groundwater quality
Oil and Gas Production	Large volume of produced, impaired water*	Produced water can impact surface and groundwater
Coal and Uranium Mining	Mining operations can generate large quantities of water	Tailings and drainage can impact surface water and ground-water
Electric Power Generation		
Thermo-electric (fossil, biomass, nuclear)	Surface water and groundwater for cooling** and scrubbing	Thermal and air emissions impact surface waters and ecology
Hydro-electric	Reservoirs lose large quantities to evaporation	Can impact water temperatures, quality, ecology
Solar PV and Wind	None during operation; minimal water use for panel and blade washing	

*Impaired water may be saline or contain contaminants

Energy Element	Connection to Water Quantity	Connection to Water Quality
Refining and Processing		
Traditional Oil and Gas Refining	Water needed to refine oil and gas	End use can impact water quality
Biofuels and Ethanol	Water for growing and refining	Refinery waste-water treatment
Synfuels and Hydrogen	Water for synthesis or steam reforming	Wastewater treatment
Energy Transportation and Storage		
Energy Pipelines	Water for hydrostatic testing	Wastewater requires treatment
Coal Slurry Pipelines	Water for slurry transport; water not returned	Final water is poor quality; requires treatment
Barge Transport of Energy	River flows and stages impact fuel delivery	Spills or accidents can impact water quality
Oil and Gas Storage Caverns	Slurry mining of caverns requires large quantities of water	Slurry disposal impacts water quality and ecology

**Includes solar and geothermal steam-electric plants

Source: U.S. Congress Report on interdependency of energy and water.

Figure 3. Connections between Energy Sector and Water Availability and Quality.

has resulted into waste streams, runoff from mining operations, and air emissions affecting the watersheds.

Main Trends and Patterns in African Water and Energy System

Africa has failed to understand the interdependence of water and energy – a reason why most of the continent is in darkness. Sub-Saharan Africa consists of 44 countries of which 13 countries are landlocked and 12 are SADC member states. Most of sub-Saharan countries have hydroelectric stations that are not functioning. Most of the landlocked countries are not well served with access to energy because they are situated in the Sahara, have political strife and conflicts such as Mali, Burkina Faso and Niger. In Western region, ECOWAS has played an important role. Cote d'Ivoire, Ghana and Nigeria is energy rich and are largest generator of electricity inspite of the fact that most of these member states are relying heavily on traditional biomass as their primary source of energy option. In Eastern Africa, the regional cooperation on electricity supply has helped current levels of energy use in the countries there. The Southern African Power Pool (SAPP) - a cooperation between large number of nations in the region, such as South Africa, Namibia, Botswana, Lesotho, Swaziland, Mozambique, Zambia, Zimbabwe and Malawi have played an active role to address access to energy in its region.

Most of the countries in Africa face chronic shortage of water due to harsh climatic conditions, relies heavily on seasonal rainfall leading to little or costly generation of electricity, does not have any domestic resource of fossil fuel for domestic use and paralysed by power loss due to bad transmission lines and power theft. This has lead to severe environmental problems such as deforestation, soil erosion, and runoff due to overuse of traditional biomass as means of the only cheap domestic energy option. Some of these problems such as seasonal rainfall are beyond the control of power

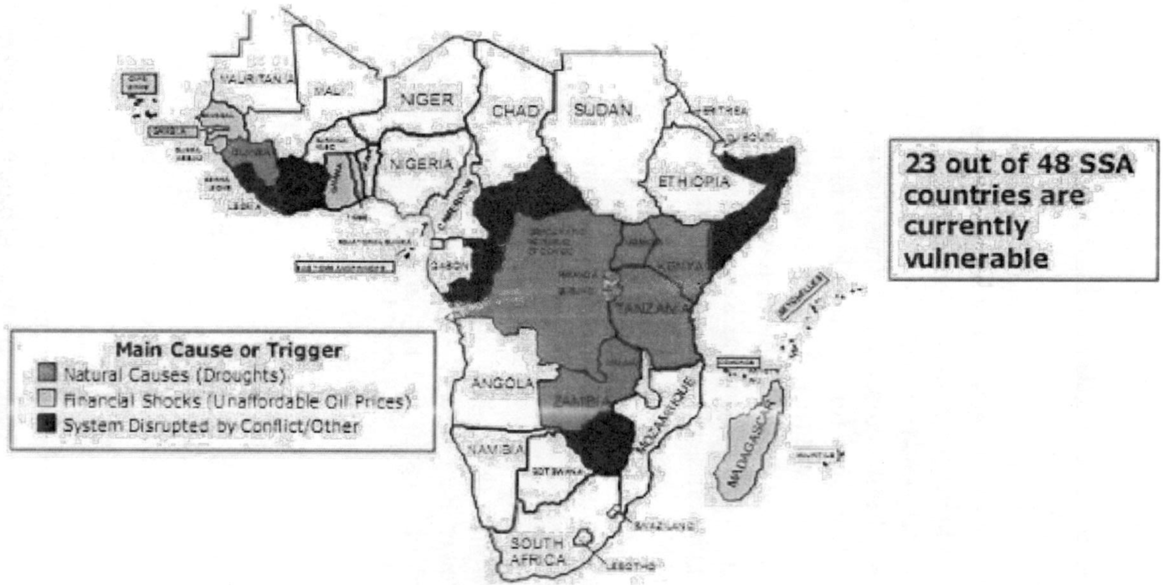

Source: Africa Energy Unit World Bank Presentation by Mac Cosgrove-Davies, Sr. Energy Specialist.

Figure 4. Areas of Ongoing or Imminent Power Shortfalls in SSA.

pools and some of these problems such as deforestation and power theft are manageable by them.

To understand the magnamity of the problems highlighted above, following are the brief descriptions[10] of the status of energy access in SSA countries. Angola, a country that is recovering from civil war - has only 15 per cent of its households having access to electricity, the main source of energy for residential use is traditional biomass. Benin, has no domestic production of natural gas, coal or oil, and imports petroleum products and relies heavily on the traditional biomass. In Botswana, a landlocked economy - much work is required in developing appropriate and efficient techniques for domestic energy options particularly renewable energy. Burkina Faso, a landlocked economy - faces chronic shortage of water impacting generation of electricity that has resulted into high costs of electricity provision. Only seven per cent of its households have access to electricity. In Burundi, a landlocked economy - around 1.5 per cent of the population has access to electricity while 86 per cent of energy consumption use is biomass, comprising of wood, charcoal and peat, and only 11 per cent is imported petrol products, and two per cent electric.

In Cameroon, not a member of the SAPP - the key constraints for a failing power sector is due to reliance on aged facilities and equipment, the effect of harsh climatic conditions, and a lack of long-term maintenance. Cape Verde, does not have any domestic resources of fossil fuels and has to import all fossil products. Central African Republic, a landlocked economy - relies heavily on wood, a supply that makes 80 per cent of the country's energy needs. In Chad, a landlocked economy - less than 2 per cent of its population has access to electricity, and 97 per cent of the population relies on wood for fuel requirements, this being a major contributor to deforestation. In Comoros, wood fuel provides about 75 per cent of the country's energy, with thermal plants supplying the remainder.

Inspite of being a member of SAPP only 19.1 per cent of the Congo DRC (Zaire), population is connected to electricity grid. Electrical consumption in Cote d'Ivoire, a WAPP member - amounts to 47 per cent supplied by hydropower while Oil and Gas accounted for a remaining 53 per cent. Countries that are connected to the Ivorian grid include Mali, Burkina Faso, Benin and Togo.

In Djibouti electricity access currently is almost non-existent in rural areas. Djibouti is characterised by having neither fossil resources (oil and gas) nor hydroelectric potential. In Equatorial Guinea renewable Energy potential is largely undeveloped. Eritrea does not produce any oil, coal or natural gas, making it dependent on petroleum product imports. Inspite of 60 per cent of the water in the Nile River is being supplied by the Blue Nile, which originates in the highlands of Ethiopia, a landlocked economy - suffers from serious shortages of electricity due to slow development of hydro. Only 47.9 per cent of Gabon's population is connected with the grid. Inspite of this, Gabon still uses traditional biomass as its residential energy leading to deforestation. Gambia faces 40 per cent power loss due to bad transmission lines and power theft. Due to this, fuel wood has remained its dominant energy resource, accounting for about 81 per cent of total energy consumed. 48.5 per cent of people in Ghana, a member of WAPP - are connected to grid; primary source of energy is use of traditional biomass that has lead to deforestation. Ghana primarily relies on hydro power from its Akosombo Dam, on the Volta River about 80 kilometers upstream from the coast. Ghana supplies Benin and Togo with the majority of their electricity.

Guinea, a member of WAPP - does not have any proven domestic reserves of coal, natural gas or oil, and is forced to import all of its fossil fuels. Traditional biomass is the primary means of energy option. No natural gas or coals are consumed in Guinea-Bissau, nor does renewable energy play any part in the country's energy mix. 66.7 per cent people use traditional fuel. Kenya is highly dependent on hydroelectricity, which is responsible for over 75 per cent of all electrical output. Energy consumption is dominated by wood fuel and other non-sustainable sources. Coal is the major energy source consumed in Lesotho, a landlocked economy supplying 87 per cent of its commercial energy needs with only two per cent of the households in the country are electrified. Wood accounts for about 90 per cent of the total energy requirement, in Liberia, due to ongoing civil unrest have caused severe fuel distribution problems and shortages.

Madagascar is currently suffering from a series of blackouts, following the depletion of fuel stocks. Being a landlocked economy 86.6 per cent of Malawi's population relies on traditional biomass as its means for energy and only 5.8 per cent of households are currently connected to electricity inspite of being a member of SAPP. Ninety per cent of energy consumed in Mali, a landlocked economy - comes from the unsustainable use of fuel-wood. Mauritania consumes significant amount of "non-commercial" (wood, biomass) energy. Mauritius is heavily dependent upon fossil fuels for power generation, transport, industry and cooking purposes. With no present domestic sources of energy besides biomass, hydropower and a little coal, Mozambique, inspite of being a SAPP member - is forced to import its fossil energy, mostly petroleum.

Namibia does not produce any coal, oil or natural gas, and is forced to import all of its petroleum. Niger, a landlocked economy - currently relies heavily on traditional fuel for households – about 77.4 per cent households use fuel wood, charcoal, bagasse (sugar cane waste) and animal and vegetable wastes as a mean for energy. In Nigeria, a member of WAPP - traditional biomass (largely wood fuels) accounts for the largest share of total energy consumption (51 per cent) followed by petroleum products (41 per cent), natural gas

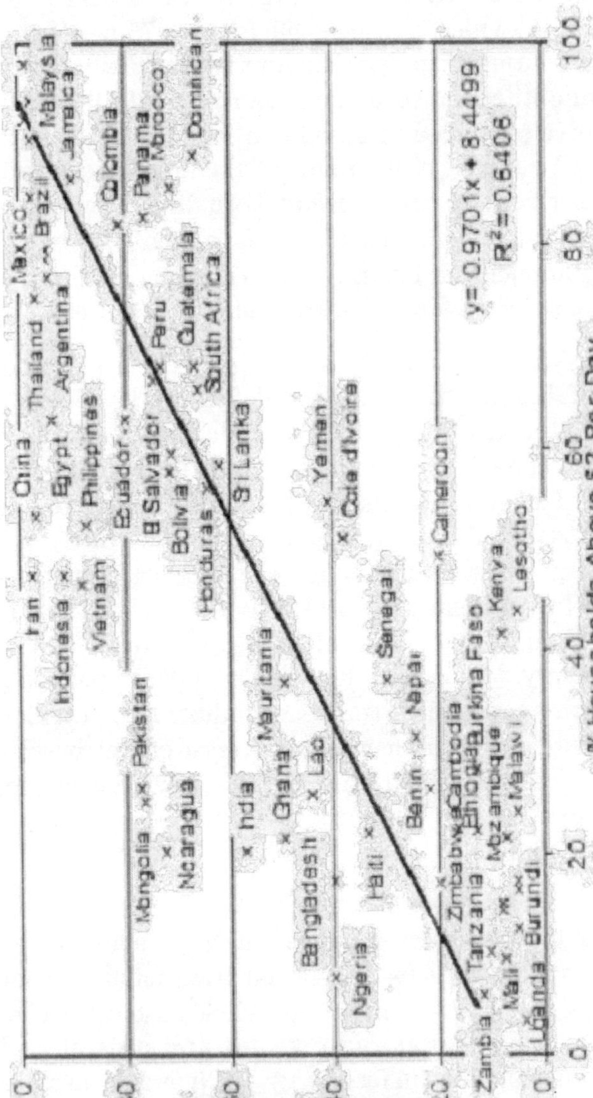

Source: Africa Energy Unit World Bank Presentation by Mac Cosgrove-Davies, Sr. Energy Specialist.

Figure 5. African Access Rates Below International Comparators.

(5.2 per cent), and electricity (3.1 per cent). Over 90 per cent of Rwandans today depend on biomass for their energy.

Approximately just over half the population of Sao Tome and Principe has access to electricity; many households are without electricity and have to rely on candlelight and kerosene lighting, as well as on biomass (firewood and charcoal) for cooking purposes. Currently only 31.4 per cent of households in Senegal are connected to electricity grid inspite of being a member of WAPP. The Seychelles does not produce any oil, natural gas or coals, and relies heavily on imported petroleum products to cover domestic demand. Only 5 per cent of the total population has access to electricity. 92 per cent of households use traditional biomass as an energy option for cooking and electricity.

In Somalia, fuel wood and agricultural residues account for 87 per cent of all energy consumed, while petroleum accounted for 11.8 per cent in the form of electricity. Highest per centage of households connected to electricity i.e. 67.1 per cent South Africa represents a wide range of investment in energy options. Both residential as well as industry energy use is dominated by wood fuel in Sudan.

Over 75 per cent of the population in Swaziland lives in rural areas, which use wood fuel as their main source of energy for cooking and heating, using inefficient stoves. The country's electricity supply has been erratic because of the national grid's heavy reliance on hydroelectric power, which in turn depends on rainfall. Combustibles like fuel wood are used for residential energy consumption in Togo, which is the main cause for deforestation.

Much of the electricity network in Uganda is poorly maintained and power cuts are frequent. Just 3-5 per cent of the population has regular access to electricity. The largest source of energy in Zambia, a landlocked economy and a SAPP member - is traditional biomass. Residential energy use consists primarily of wood fuel. Zimbabwe, a landlocked economy and a member of SAPP - uses biomass which accounts for 50 per cent of the energy used, while coal and electricity account for 13 per cent and 12 per cent respectively.

Lawrence Berkeley Laboratory Case Study: Wind Power for Rural Energy Services in Eritrea using Desalinisation[11]

This is brief information of Lawrence Berkeley laboratories desalination project in Eritrea to address unavailability of clean water and access to energy. This project was motivated by proximity of salt water to a potentially sizable energy source wind power; to desalinate seawater or brackish groundwater for village use. The study identified that energy requirement for brackish water desalination range from 1 to 3 kWh/m^3, while requirements for seawater desalination can range from 5 to 15 kWh/m^3. (Loupasis, 2002)

To establish a renewable energy technology accurate sizing system; in this case wind power, an important first step laboratory took was obtaining an accurate value of the energy available over the course of a year to meet a given level of demand. Scientist faced problems when pairing wind turbines with desalination systems due the intermittency of the wind source in Eritrea where only 316 of its 2500 villages in areas have favourable wind regimes. This aspect of the technology the scientist agree does not Technical Advisor for Eritrea Wind Energy Applications Project Prepared by Douglas M. Bushey, Lawrence Berkeley National Laboratory, UC suit the commercial aspect of generating power, particularly when local demand is not met in periods of low wind speeds.

The scientist rectified this problem by installing a large enough water storage tank to store water through these periods. But the most challenging issue they faced was that wind intermittency raised incompatibility of many desalination units due to discontinuous energy supply. Some systems waste a large amount of energy when the power shuts off in the mid of the process; while others may not sufficiently desalinate water at low or variable energy supply rates.

This, scientist believe can be a deciding factor for implementing renewable energy options to choose whether or not to charge for water for the community. Some argue that drinking water can be provided to the largest number of people most quickly by charging for water, so that investments in water provision pay back and generate capital to invest in new projects. The government could also invest in these projects, as an alternative and generate capital by providing water and electricity at a subsidised rate for return of investment.

Financing options for achieving the water and energy objectives in Africa[12]:

This section includes brief highlights of investments done by the international community in Africa. GEF has currently invested $32.33 million with a co-financing of $46.38 million dollars in 44 countries through its 274 projects that addresses energy efficiency, land degradation, trans-boundary systems and water management in one form or the other within the continent. (ANNEX II for more details regarding these projects)

Within this context following projects have been initiated or completed under GEF/UNDP's supervision: 1. Participatory Management of Natural Forests and Village Reforestation to reduce Carbon Emissions in Benin. This is very important for this country, because Benin has no domestic production of natural gas, coal or oil and relies heavily on biomass as its primary means of energy option (operationally completed in December 2008) ; 2. RE-Based Rural Electrification Programme is being implemented in Botswana whose economy is having the problem of low uptake of electricity connections caused by the high up-front costs (under implementation); 3. Wind Energy Applications in Eritrea has assisted in addressing those 80 per cent of the Eritrean population without access to electricity (under implementation); 4. Renewable Energy-based electricity project used for generating electricity for Rural, Social and Economic Development in Ghana addresses 50 per cent of its population without electricity (operationally complete on 16th June 2001); 5. Removal of barriers for achieving energy conservation and energy efficiency in small and medium scale enterprises in Kenya has addressed energy consumption that is dominated by wood fuel and other non-sustainable sources (project closed on 13th December 2006); 6. RE-based rural electrification project in Lesotho addresses the needs of 98 per cent of the households in the country that are not electrified. This is mostly due to huge connection costs charged by utilities (under implementation); 7. Decentralised wind electric power for social and economic development (Alizes Electriques) in Mauritania is addressing tremendous untapped potential to provide electricity in the country (project closed on 31st December 1998); 8. A project for removal of Barriers to Energy Efficiency and Energy Conservation in Buildings in Mauritius was important for the economy that relied heavily on fossil fuels (under implementation); 9. Barrier removal project meant to identify and remove barriers in Namibia Renewable Energy Programme (NAMREP) for an economy that relies heavily on traditional biomass (under implementation); 10. South Africa Wind Energy Programme (SAWEP) – an economy that represents wide range of investment in energy options (under implementation); 11. An investment in a biogas plant for Africa: electricity, fuel and fertiliser from municipal and industrial waste in Tanzania: (under implementation).

World Bank's Africa Energy Unit[13] believes that meeting Africa's Energy Challenges requires a radical scaling-up of access to electricity for growth, by increasing coverage for household (>35 per cent on average), productive enterprises

(100 per cent mostly in urban and peri-urban areas) and rural areas close to the grid.

World Bank Africa Energy Unit, believes that for powering the MDGs it has to provide electricity access to 75 per cent of schools, clinics, community centres and local administration; by using grid extensions with least cost and decentralised solutions such as solar PV, independent grids, small renewables generation. World Bank believes in meeting basic needs meant (1) >50 per cent households equipped with at least one modern, affordable light, (2) >75 per cent households use improved cook stove or LPG and sustainable fuel-wood supply.

In the 90s World Bank through its PRSPs review did number of reforms with a focus on energy sector such as domestic cooking fuels, modern fuels such as LPG, renewable energy, and policies to facilitate energy access to the poorest. Although it failed to look into the interdependency factor of other sectors such as water, land, agriculture etc.

Within the continent, ECOWAS member states have come up with an ambitious objective to obtain access to energy services by 2015. This objective includes, "I. supplying 36 million more households and 49,000 more localities with access to energy services by 2015. II. 60 per cent of people living in rural areas will have access to productive energy services in villages, in particular motive power to boost the productivity of economic activities; III. Connecting 66 per cent of the population, or 214 million people, will have access to an individual electricity supply."[14]

"ECOWAS member states, have proposed a total estimated cost of activities to be carried out in the region as 248.7m $ (implementation of the issues for action) over a 10-year period."

The total estimated cost is 248.7 million $ over a 10-year period and can be broken down as follows amongst the four lines of action:

Line of action 1: capacity building . . . . 83.1 M$

Line of action 2: support to funds mobilisation . . . . 121.2 M$

Line of action 3: promotion and experience dissemination . . . . 15.6 M$

Line of action 4: promotion of local equipment production .... 12 M$

Preparatory activities and operation of the agency over 10 years . . . .16.8 M$

The average cost of connecting a household is estimated at 350M$.

"The massive recourse to traditional fuels – 67 per cent dependency on biomass in Africa, and 80 per cent in ECOWAS countries - is because of people's low purchasing power, which prohibits them from affording modern fuels: their low income means they consume much more energy per unit of added value than in developed countries. This over-consumption takes a toll on the environment (soil erosion, desertification, etc.). Therefore, the absence of modern fuels propels the poverty spiral further downward (ECOWAS 2005 policy document)".

The ECOWAS objective for 2015 is for all of the population to have access to a modern or improved cooking service, through:

"• Access to modern fuels – which requires, in the case of LPG, that households buy a gas stove and canister.

• Improved biomass stoves, in conjunction with the construction of chimneys to reduce in-door air pollution. Where biomass is used, biomass production must be sustainable, using sustainable energy crops. This entails, as we saw earlier, carrying out in-depth reform of the forestry and rural sectors.

In order to reach LPG access targets, 30 million stoves must be supplied over the next ten years."[15]

ECOWAS adopted a resolution to create a collective energy autonomy called West African Power Pool system in a view of boosting interconnection capacity. It has come out with a total of $57.3 billion for ten years to achieve its objective.

The reason for disappointing results around implementing ECOWAS or SADC policy outcomes has been due to decreasing investment in ODA and plummeting private investments in Africa. This has lead to energy ministries and industries forced to invest in costly private alternative energy sources. Under capitalisation has lead to decrease in infrastructure development for energy and water sector. Energy reforms failed in Africa due to under capitalisation in infrastructure development sector.

It proposes the following for achieving energy access in Western Africa:

- "17.5 billion dollars over ten years for investment in equipment needed for access to modern energy services including studies and accompanying measures;

- 34.6 billion dollars over ten years for energy costs, or 3.46 billion dollars per year; the figure is an overall total that includes depreciation of generation and transmission costs: given the amount required, it goes without saying that consumers will not be able to foot all of this bill, therefore states will have to provide subsidies, with each one to decide how much;

- The overall cost of three programmes is around 16 dollars per inhabitant per year;

- The total annual cost for programmes is estimated at US$ 5.2 Billion, around 4.3 per cent of regional GDP."[16]

Applying ECOWAS baseline calculation for investing in infrastructure equipments to access energy and water for

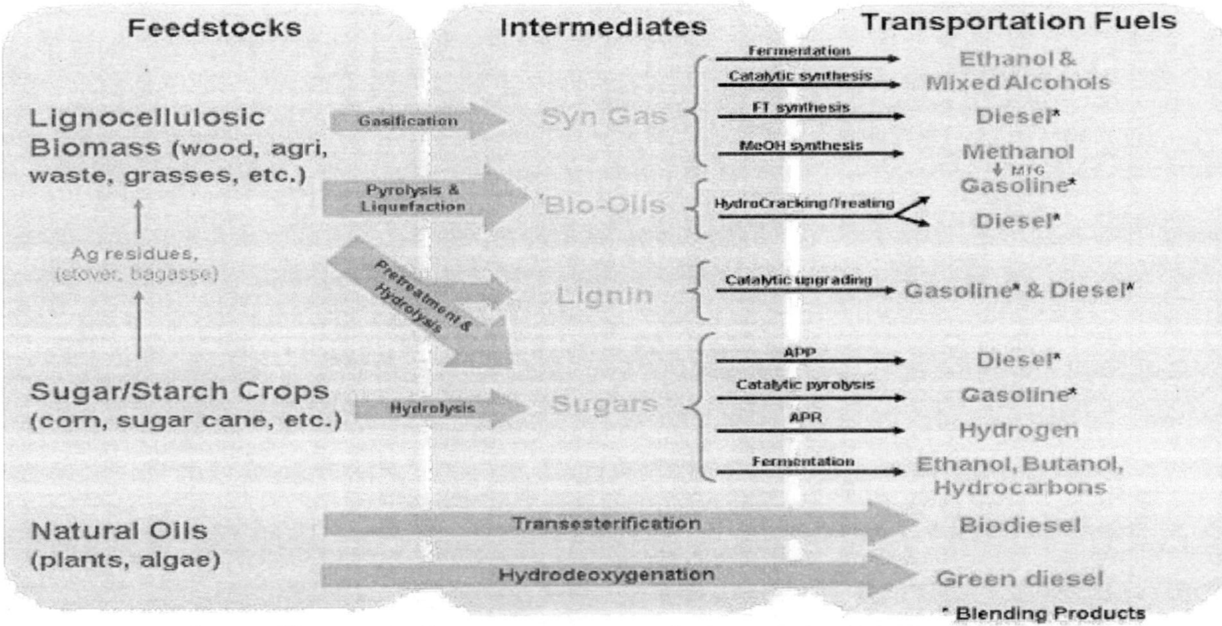

Source: National Renewable Energy Laboratory's, Presentation by Director Dan E. Arvizu.

Figure 6. Examples of Pathways to Biofuels.

entire Africa, the total cost comes to an overwhelming $2120.10 Billion over ten years.

ECOWAS member states have recommended investigation of the market potential for bio-fuels and the conditions for developing industrial crops on an ad hoc basis. The continent, could perhaps consider investigating the potential of non-food biomass feedstock – the next generation biofuels which will not compete with the food, land and water resources of the region. US National Renewable Energy Laboratory suggests a concentrated research on Ligno-cellulosic biomass "a complex matrix of cellulose, hemi-cellulose, and ligniniv polymers that differ from the easily hydrolysable starch from grains and easy-to-process juice in sugar cane. Because of the compositional and morphological complexity of the polymers, technical challenges exist for their conversion into ethanol or other biofuels."[17]

Conclusions

While energy is a key infrastructure engine of development, Africa continues lagging behind other regions in this area. The core energy challenges in Africa are access to energy services and security of water supply. This is primarily because Africa has not taken stock of an integrated approach of the impact of appropriate water policies and regulations on energy supplies while those of energy policies and regulations on water supplies and demands.

It became clear that the landlocked and sparsely populated economies are not particularly well served with energy, mainly because they are relatively poor and are at least partially situated in the Sahara. Energy development is also very limited in the small coastal countries of Liberia, Sierra Leone, Guinea and Guinea-Bissau because of small economies, political strife and conflicts.

In East Africa, regional cooperation model on electricity supply has helped sustain current levels of energy use in the countries there, where most of the electricity supply is derived from the Nalubaale (formerly Owen Falls) hydro scheme in Uganda. More recently the East African Community, consisting of Kenya, Uganda and Tanzania, has initiated the preparation of an East African Power Master Plan with the assistance of the World Bank and the Swedish International Development Agency. There will be an increased supply of electricity to Kenya once the planned Bujagali scheme is commissioned in 2011 by latest estimates. Kenya also hosts Ol Karia Power Station which is the continent's biggest geothermal power-generating plant.

In southern Africa, the South African Power Pool (SAPP) model has helped to network a large number of nations in the region. These include South Africa, Namibia, Botswana, Lesotho, Swaziland, Mozambique, Zambia, Zimbabwe and Malawi.[18] On similar grounds West African Power Pool (WAPP) has made a difference to nations in the western region.

GEF has accepted the fact that there is lack of awareness about the potential benefits of developing the renewable energy and energy efficiency markets in Africa to effectively address the energy challenges, and subsequently, about their social and economic benefits in the region.

The strategic energy investment of the GEF programmes specifically focuses on practical interventions on the ground that demonstrate the technical and economic viability of promising renewable energy and efficient energy technologies and measures, and promote private sector involvement in stimulating energy markets in 44 countries of the region. Although it is time for the GEF to scale up from the demonstration projects to large scale projects with World Bank.

The World Bank believes that their consistent efforts to scale-up of access, to ensure affordable, reliable, and sustainable supplies of energy services, is part of their broader efforts to stimulate growth, reduce poverty, and promote inclusiveness in Africa. Energy access along with good governance and efficient regulatory framework they believe would be a certain pathway for productive steady growth in the region.

To address this, African countries have been investigating in ways to strengthen regional integration by pooling good practices, experience sharing, adopting a common regional information system and developing trans-boundary co-operation, that foster's peace, development and building local capacities.

The study looked at why it has been difficult to implement renewable energy options in the continent and identified four barriers that offsets there important role. These barriers are: (i) high initial technology costs; (ii) local capacity to maintain the infrastructure; (iii) delivery infrastructure (supplier-dealer chains, service arrangements) which lacks the volume to achieve economies of scale; and (iv) lack of credit to system users (as a consequence of the high first cost of the system). Therefore, in order to recognise the full potential of off-grid/ de-centralised rural electrification, an effective integrated water and energy policy is very important. This might also address the environmental problems in the continent.

In conclusion, the fact remains clear that the delivery of electricity or energy services will not automatically lead to economic development in Africa, this implies that some additional intervention or efforts are needed in areas such as governance, enabling environment, integrated policies, efficient management, and effective regulatory framework. Findings indicate that Africa is not resilient towards ecosystem, economy and bureaucracy to address these challenges.

Recommendations:

- Recommitting to the creation of a new International Water Agency to monitor the water levels, local capacity to adapt to changing weather conditions, river basin management, water conflicts, and addressing the adaptation of water resources management for local capacity building in economies that are impacted by depletion of the water level.

- Most of the African states use hydropower to generate electricity and not geothermal, solar or wind power. It is therefore necessary to understand this trend of use of renewables which might be an indicator to address the energy security within the continent.
- Consideration should be given to increased competition for water resources both domestically and trans-boundary, by identifying ways to reduce these conflicts through coordinated infrastructure development.
- Global Water Partnership (GWP) should invest in addressing its success around strengthening the in-house technical capacity building within each African state in sustainable water resources and drought management.
- GWP should consider evaluating its ground services in the form of technical assistance policy from the context whether the African states have been able to self-sustain its water and sanitation issues at domestic, national and regional level.
- International community could consider assisting the African economies to develop domestic water and energy policies that are pro-development and transparent governance.
- Priority needs to be given to address capacity development to mainstream, integrate and introduce renewable energy within the local community for sustained use of these off grid energy options.
- Creating an enabling environment for the national governments to develop budgets investing in energy access through decentralised energy services systems.

- Investigating the appropriateness of oil revenue-funded economic development plans with PRSPs and other international development initiatives to address the prioritisation of their implementation for an underdeveloped economy.
- Giving priority to local capacity building to address lack of multi-sectoral institutional structures, the only way to perhaps overcome the challenges posed by inadequate regulatory framework and political will.
- Supporting ECOWAS recommendation of creating a very important regional agency for widening access to energy services; that will assist in collecting data of the member states regarding energy services, infrastructure systems, and relevant players in the field.
- Africa should make an effort to come up with its own water-energy research roadmap to address steadily decreasing water resource and sustain a reliable energy future.
- Consider investing into conservation funding to achieve conservation pricing, incentives, and savings.

Country Assessment Summary

This section of country assessment was done to understand the inter-linkages between a nations water conflicts, its energy challenges and how it corroborates to its economic as well as national security.

This assessment studied every countries demography; blue, green and grey water footprints i.e. rivers, basins, estuaries that flow throughout the country accountable for thriving the economy; its conflicts arisen due water shortages; energy options used to scale up the economy, and the challenges it faces w.r.t. access to energy.

It accomplished the task by studying 1) the U.N. & World Bank reports, 2) National Energy & Water Policies, 3) IEA projects, 4) the World Water series/ policies, 5) projects funded by multilateral agencies such as UNDP & GEF, and 6) NEPAD/ECOWAS/SADC/African Union regional reports.

The assessment concludes that the lack of national energy policies integrated with water management and conflicts in Africa have severally crippled its infrastructure and havn't been able to upgrade the existing decades old system due to lack of resources and investments, further worsening the access to energy crisis. This "quintessential paradox" in Africa – the continent that has diamonds, water, oil, rivers, forests, and flora-fauna havn't been able to be used by the Africans themselves, for their economic development and strengthening their national security.

Inspite of Africa having sufficient underground aquifiers, because of ineffective water policies whose implementation is further hindered by droughts, unaffordable oil prices, disruption of law & order system due to conflicts, overall citizens have faced shortages of water access.

The leaders that have gotten its economy right by placing emphasis on effective energy budgets, water management strategies and mainstreaming renewable off-grid technologies into the communities are South Africa, Kenya, Rwanda, Nigeria, Ghana, Tanzania, Uganda, Gabon, Cameroon, Botswana, Senegal, Algeria, Zambia, Ethiopia, Morocco, Mozambique, Sao Tome & Principe, Liberia, Malawi & Mauritius.

Brief information regarding population, rivers, annual renewable water resources, water conflicts, energy opportunities and challenges of each countries in Africa

African State	*Population (million) /GDP Billion $*	*River/Basin*	*Annual Renewable Water[19] Resources[20] (km³/yr)*	*Water Conflicts and Challenges*	*Energy opportunities*		*Energy challenges*
					Renewable source of Energy for electricity production	Oil Production (KTOE)	
Algeria	33.25/ 69.70	Lake Chad, Niger	14.3		Hydro	90942	1) Imports coal and crude petroleum.[21] 2) electricity production ought to double and rise from 24,400 GWH currently to around 46,000 GWH in 2010[22] 3) Undeveloped infrastructure

(Contd.)

(Contd. Annex-1)

African State	*Population (million) / GDP Billion $*	*River / Basin*	*Annual Renewable Water[19] Resources[20] (km³ / yr)*	*Water Conflicts and Challenges*	*Energy opportunities*		*Energy challenges*
					Renewable source of Energy for electricity production	Oil Production (KTOE)	
Angola	15.94/ 14.94	Cunene; Zaire; Okavango Makgadikgadi, Zambezi	184.0	Local water demands are low as the Cunene river lies in sparsely populated area. Opposition to the hydro-electrical project in the Epupa scheme was voiced; Zaire, forms the border between the DRC an Angola. Other countries in the basin are	Hydro	63623	1) Civilian attacks has left country's energy and water infrastructure in decay 2) Imports petroleum products[23] 3) Only 15% of the country has access to electricity, as the current capacity is 586 megawatts for 13 million inhabitants[29];

(Contd.)

(Contd. Annex-1)

African State	*Population (million) /GDP Billion $*	*River/ Basin*	*Annual Renewable Water[19] Resources[20] (km³/yr)*	*Water Conflicts and Challenges*	*Energy opportunities*		*Energy challenges*
					Renewable source of Energy for electricity production	*Oil Production (KTOE)*	
				Cameroon, the Central African Republic and the Republic of Congo. Potential conflict could arise out of its vast surplus of water (ten times that of the Zambezi) The Inga Rapids have the largest single hydroelectric energy potential in the world.			4) Of the country's six hydroelectric stations, only three are currently functioning[30]; 5) Extensive repair and investment needed in both water processing and in distribution networks[31]

(Contd.)

(Contd. Annex-1)

African State	Population (million) /GDP Billion $	River/ Basin	Annual Renewable Water[19] Resources[20] (km³/yr)	Water Conflicts and Challenges	Energy opportunities		Energy challenges
					Renewable source of Energy for electricity production	Oil Production (KTOE)	
				Cameroon, the Central African Republic and the Republic of Congo. Potential conflict could arise out of its vast surplus of water (ten times that of the Zambezi) The Inga Rapids have the largest single hydroelectric energy potential in the world.			4) Of the country's six hydroelectric stations, only three are currently functioning[30]; 5) Extensive repair and investment needed in both water processing and in distribution networks[31]

(Contd.)

(Contd. Annex-1)

African State	*Population (million) /GDP Billion $*	*River/ Basin*	*Annual Renewable Water[19] Resources[20] (km³/yr)*	*Water Conflicts and Challenges*	*Energy opportunities*		*Energy challenges*
					Renewable source of Energy for electricity production	*Oil Production (KTOE)*	
				100 bodies were found in four drinking water wells in central Angola.[23] In September 1998, fierce fighting between UNITA and Angolan government forces broke out at Gove Dam on the Kunene River for control of the installation.[24] Cuban and Angolan forces launch an attack on Calueque Dam via land and then air. Considerable damage inflicted on dam wall; power supply to dam cut. Water pipeline to Owamboland cut and destroyed.[25]			4)The main source of energy for residential use is traditional biomass.[32]

(Contd.)

(Contd. Annex-1)

African State	Population (million) /GDP Billion $	River/ Basin	Annual Renewable Water[19] Resources[20] (km^3/yr)	Water Conflicts and Challenges	Energy opportunities		Energy challenges
					Renewable source of Energy for electricity production	Oil Production (KTOE)	
				South African troops move into Angola to occupy and defend the Ruacana hydropower complex, including the Govè Dam on the Kunene River. Goal is to take possession of and defend the water resources of southwestern Africa and Namibia.[27]			

(Contd.)

(Contd. Annex-1)

African State	Population (million) /GDP Billion $	River/ Basin	Annual Renewable Water[19] Resources[20] (km^3/yr)	Water Conflicts and Challenges	Energy opportunities		Energy challenges
					Renewable source of Energy for electricity production	Oil Production (KTOE)	
Benin	2.7/5.92	Niger, Volta	25.8		Combustible renewables and waste	0	1) Imports petroleum products 2) Conventional energy is expensive and difficult to find 3) Huge demand for solar energy although resources are not available[33] 4) Benin has no domestic production of natural gas, coal or oil.[34]

(Contd.)

(Contd. Annex-1)

African State	*Population (million) /GDP Billion $*	*River/ Basin*	*Annual Renewable Water[13] Resources[20] (km^3/yr)*	*Water Conflicts and Challenges*	*Energy opportunities*		*Energy challenges*
					Renewable source of Energy for electricity production	*Oil Production (KTOE)*	
Botswana	1.86/8.39	Limpopo, Okavango Makgadikgadi, Orange, Zambezi,	14.7	Conflict w/ upstream Namibia over use of water from Okavango Makgadikgadi to sustain its lucrative ecotourism, potential conflict w/ Lesotho Highland Water Project involving Orange river[35];	Combustible renewables and waste	0	Much work is required in developing appropriate and efficient coal burning appliances and equipment for both institutional and domestic coal use. Though the construction of a second power station is being deferred, the country is

(Contd.)

(Contd. Annex-1)

African State	Population (million) /GDP Billion $	River/ Basin	Annual Renewable Water[19] Resources[20] (km³/yr)	Water Conflicts and Challenges	Energy opportunities		Energy challenges
					Renewable source of Energy for electricity production	Oil Production (KTOE)	
				The disputed ownership of Sedudu / Kasikili island in the Chobe River between Namibia and Botswana is an example for causing a water-related conflict.[36]			challenged to put this abundant locally available resource to use to save foreign exchange[37] the problem of low uptake of electricity connection caused by the high up-front costs[38] Solar photovoltaic electrification is taking off in Botswana Solar water heating is one area needing to be fully addressed to benefit the country on a much larger scale

(Contd.)

(Contd. Annex-1)

African State	*Population (million) /GDP Billion $*	*River/ Basin*	*Annual Renewable Water[19] Resources[20] (km³/yr)*	*Water Conflicts and Challenges*	*Energy opportunities*		*Energy challenges*
					Renewable source of Energy for electricity production	*Oil Production (KTOE)*	
							Wind energy applications are at present limited to water pumping with few scattered wind pumps across the country due mainly to low average wind speeds Botswana's large herds of cattle, slaughtering points can provide substantial livestock waste for biogas production for institutional energy use.[39]

(Contd.)

(Contd. Annex-1)

African State	Population (million) /GDP Billion $	River/ Basin	Annual Renewable Water[19] Resources[20] (km^3/yr)	Water Conflicts and Challenges	Energy opportunities		Energy challenges
					Renewable source of Energy for electricity production	Oil Production (KTOE)	
Burkina Faso	14.33/	Niger, Volta,	17.5	Declining rainfall has led to growing fights between animal herders and farmers with competing needs. In August, 2007 people were forced to flee their homes by fighting in Zounweogo province.[40]			1) Chronic shortage of water impacts generation of electricity. Resulting into high costs of electricity provision. This has resulted into importing electricity from Coted'Ivoire.[41] 2) Burkina Faso has no national transmission network and the distribution network is limited. Only 7% of the country has access to electricity and electricity is expensive.[42]

(Contd.)

(Contd. Annex-1)

African State	*Population (million) /GDP Billion $*	*River/ Basin*	*Annual Renewable Water[19] Resources[20] (km^3/yr)*	*Water Conflicts and Challenges*	*Energy opportunities*		*Energy challenges*
					Renewable source of Energy for electricity production	*Oil Production (KTOE)*	
							3) 91.7% percent Population dependant on fuel wood, charcoal, bagasse, animal and vegetable waste.[43]
Burundi	8.40/	Zaire, Nile,	3.6				Burundi has no resources of coal, natural gas, or petroleum meaning that electricity production in the country was very problematic. Hydroelectric power stations now account for most of the country's electricity production as they do not have the natural reserves to produce it.[44]

(Contd.)

(Contd. Annex-1)

African State	*Population (million) /GDP Billion $*	*River/ Basin*	*Annual Renewable Water[19] Resources[20] (km^3/yr)*	*Water Conflicts and Challenges*	*Energy opportunities*		*Energy challenges*
					Renewable source of Energy for electricity production	*Oil Production (KTOE)*	
							Burundi's energy consumption is biomass, comprising of wood, charcoal and peat, and only 11% is imported petrol products, and 2% electric. Only 1.5% of the population has access to electricity, over 90% of which is consumed in Bujumbura. Rebel attacks through out 2003 and the still not fully implemented cease fire accords and agreements between the

(Contd.)

(Contd. Annex-1)

African State	*Population (million) /GDP Billion $*	*River/ Basin*	*Annual Renewable Water[19] Resources[20] (km³/yr)*	*Water Conflicts and Challenges*	*Energy opportunities*		*Energy challenges*
					Renewable source of Energy for electricity production	*Oil Production (KTOE)*	
							various warring sides in Burundi have resulted in several power disruptions to the Burundian capital and have prevented any hydroelectric energy development in the countryside.[45] There is no private ownership of hydro-plants. The government is promoting peat production and is fostering the development of renewable energy resources, such as solar electricity and biogas due to scarce firewood and charcoal for domestic use.[46]

(Contd.)

(Contd. Annex-1)

African State	*Population (million) /GDP Billion $*	*River/ Basin*	*Annual Renewable Water[19] Resources[20] (km³/yr)*	*Water Conflicts and Challenges*	*Energy opportunities*		*Energy challenges*
					Renewable source of Energy for electricity production	*Oil Production (KTOE)*	
Cameroon	18.18/12.53	Zaire, Lake Chad, Niger,	285.5		Hydro	6119	Energy Sector centralised with Government.[47] Shortfall is due to a lack of efficient maintenance the use of aged facilities and equipment, and the effect of variable climatic conditions.[48] The performance of the power sector since 2001 has been disappointing, as severe droughts have limited the supply of water, which is the source of 90 percent of the electric power produced in Cameroon. To guard against outages in the dry season, Cameroon must decrease its dependency on hydropower through

(Contd.)

(Contd. Annex-1)

African State	Population (million) /GDP Billion $	River/ Basin	Annual Renewable Water[19] Resources[20] (km³/yr)	Water Conflicts and Challenges	Energy oppor .unities		Energy challenges
					Renewable source of Energy for electricity production	Oil Production (KTOE)	
Canary Islands (Spain)							diversification of its power supply.[49] Key constraints for a failing power sector is due to reliance on aged facilities & equipment, the effect of harsh climatic conditions, and a lack of long term maintenance.[50]

(Contd.)

(Contd. Annex-1)

African State	Population (million) /GDP Billion $	River/ Basin	Annual Renewable Water[19] Resources[20] (km³/yr)	Water Conflicts and Challenges	Energy opportunities		Energy challenges
					Renewable source of Energy for electricity production	Oil Production (KTOE)	
Cape Verde	0.423/		0.3				Cape Verde does not have any domestic resources of fossil fuels and has to import all fossil products, mainly liquefied petroleum gas.[51]
Central African Republic	4.37/	Zaire, Lake Chad,	144.4				Wood supplies 80% of the country's energy needs.[52] The Central African Republic imports fossil fuels from neighboring countries.[53]

(Contd.)

(Contd. Annex-1)

African State	Population (million) /GDP Billion $	River/ Basin	Annual Renewable Water[19] Resources[20] (km^3/yr)	Water Conflicts and Challenges	Energy opportunities		Energy challenges
					Renewable source of Energy for electricity production	Oil Production (KTOE)	
Ceuta (Spain)							
Chad	9.89/	Lake Chad, Niger,	43.0	Inadequate supplies of potable water; and improper waste disposal system in rural areas has contributed to soil and water pollution[54]			Less than 2% of Chad's population have access to electricity, and 97% of the population relies on wood for fuel requirements, this being a major contributor to deforestation.[55]
				In the 1960s four countries shared Lake Chad: Niger, Nigeria, Chad, and Cameroon. By 2001 it was shared by only two of those, Chad and Cameroon. The			The high cost of importing petroleum to fuel power generation makes Chad's electricity prices among the highest in the world.[56]

(Contd.)

(Contd. Annex-1)

African State	Population (million) /GDP Billion $	River/ Basin	Annual Renewable Water[19] Resources[20] (km^3/yr)	Water Conflicts and Challenges	Energy opportunities		Energy challenges
					Renewable source of Energy for electricity production	Oil Production (KTOE)	
				1980s Nigerian fishermen began migrating into Cameroon, following the retreating waters, and in the mid 1990s more than thirty Lake Chad Basin villages founded by Nigerians were counted in Cameroon. Tension arose between Nigeria and Cameroon when the former established state control and public services in these Nigerian-populated villages, including military and police forces, schools, and health centers.			

(Contd.)

(Contd. Annex-1)

African State	Population (million) /GDP Billion $	River/ Basin	Annual Renewable Water[19] Resources[20] (km³/yr)	Water Conflicts and Challenges	Energy opportunities		Energy challenges
					Renewable source of Energy for electricity production	Oil Production (KTOE)	
Comoros			1.2		Hydro		Comoros depends on imports for its fossil fuels. Wood fuel provides about 75% of the country's energy, with thermal plants supplying the remainder.[57]

(Contd.)

(Contd. Annex-1)

African State	*Population (million) /GDP Billion $*	*River/ Basin*	*Annual Renewable Water[19] Resources[20] (km³/yr)*	*Water Conflicts and Challenges*	*Energy opportunities*		*Energy challenges*
					Renewable source of Energy for electricity production	*Oil Production (KTOE)*	
Côte d'Ivoire (*Republic of Côte d'Ivoire*	18.91/	Niger, Volta	81	Declining rainfall has led to growing fights between animal herders and farmers with competing needs.	Hydro	2109	Despite large crude oil reserves, Côte d'Ivoire still imports crude oil to be processed in refineries and exported

(Contd.)

(Contd. Annex-1)

African State	Population (million) /GDP Billion $	River/ Basin	Annual Renewable Water[19] Resources[20] (km³/yr)	Water Conflicts and Challenges	Energy opportunities		Energy challenges
					Renewable source of Energy for electricity production	Oil Production (KTOE)	
				In August, 2007 people were forced to flee their homes by fighting in Zounweogo province.[58]			as petroleum. Electrical consumption amounts to 47 per cent supplied by hydropower while Oil and gas accounted for a remaining 53 per cent of the supply.[59] The use of gas-fired electricity plants has turned the country into a regional exporter of electricity. Privatization of the energy and water sectors and an increased emphasis on thermal production are important policies being encouraged by the government of Cote d'Ivoire, so renewable energy is not being actively promoted.[60]

(Contd.)

(Contd. Annex-1)

African State	Population (million) /GDP Billion $	River/ Basin	Annual Renewable Water[19] Resources[20] (km³/yr)	Water Conflicts and Challenges	Energy opportunities		Energy challenges
					Renewable source of Energy for electricity production	Oil Production (KTOE)	
Congo	3.69/4.23	Zaire,	832.0		Hydro	12972	

(Contd.)

(Contd. Annex-1)

African State	*Population (million) /GDP Billion $*	*River/ Basin*	*Annual Renewable Water[19] Resources[20] (km³/yr)*	*Water Conflicts and Challenges*	*Energy opportunities*		*Energy challenges*
					Renewable source of Energy for electricity production	*Oil Production (KTOE)*	
Dem. Rep. Congo (*Zaire*)	60.64/5.51	Zaire, Nile, Zambezi	1283	Zaire, forms the border between the DRC and Angola. Other countries in the basin are Cameroon, the Central African Republic and the Republic of Congo. Potential conflict could arise out of its vast surplus of water (ten times that of the Zambezi) The Inga Rapids have the largest single hydroelectric energy potential in the world.	Hydro	991	Over 90% of the energy consumed in DRC comes from traditional biomass, mainly used by the 65% rural population, as well as by the industry sector. In past Congo has Exporter approximately 200 GWh of electricity to neighboring countries. Of the total electrical energy produced 93.75% comes from hydropower production. Only 19.1% of the

(Contd.)

(Contd. Annex-1)

African State	*Population (million) /GDP Billion $*	*River/ Basin*	*Annual Renewable Water[19] Resources[20] (km^3/yr)*	*Water Conflicts and Challenges*	*Energy opportunities*		*Energy challenges*
					Renewable source of Energy for electricity production	*Oil Production (KTOE)*	
				Attacks on Inga Dam during efforts to topple President Kabila. This lead to disruption of electricity supplies from Inga Dam and water supplies to Kinshasa.[61]			population is connected to electricity grid.[62]
Djibouti	0.497/		0.3	Inadequate supply of potable water.			Djibouti is characterized by having neither fossil resources (oil and gas) nor hydroelectric potential. The country therefore currently has a strong dependency on imported oil products - mostly imported from nearby Saudi Arabia and Dubai – which leads to a high cost of energy.

(Contd.)

(Contd. Annex-1)

African State	Population (million) /GDP Billion $	River/ Basin	Annual Renewable Water[19] Resources[20] (km^3/yr)	Water Conflicts and Challenges	Energy opportunities		Energy challenges
					Renewable source of Energy for electricity production	Oil Production (KTOE)	
							To continue to provide power to rural residents, the government, with the help of a grant from a number of Arab financial institutions, is installing solar and wind capacity. The primary goal of the project is to replace old diesel powered rural water pumps with new ones powered by renewable resources, but excess energy will be used for electrification.[63] Electricity access currently is almost non-existent in rural areas. Djibouti shows good potential for renewable energy development, especially in

(Contd.)

(Contd. Annex-1)

African State	Population (million) /GDP Billion $	River/ Basin	Annual Renewable Water[19] Resources[20] (km^3/yr)	Water Conflicts and Challenges	Energy opportunities		Energy challenges
					Renewable source of Energy for electricity production	Oil Production (KTOE)	
							geothermal, solar and wind.[54]
Egypt	74.17/ 127.85	Nile,	86.8	Situated downstream from a long line of countries with access to the Nile, water occupies a central plank of Egypt's foreign policy and national security stance[55] Hostilities broke between Egypt and Sudan over Egypt's planned construction of the High Dam at Aswan; Long standing tensions over the Nile, especially the Blue Nile, originating in	Hydro, Geothermal, Solar etc	33020	Egypt produces practically all the energy it consumes. Electricity is generated from oil-powered stations and hydro-powered turbines on the Nile, especially those of the Aswan High Dam. Oil and, increasingly, gas and their derivatives supplied other types of

(Contd.)

(Contd. Annex-1)

African State	Population (million) /GDP Billion $	River/ Basin	Annual Renewable Water[19] Resources[20] (km^3/yr)	Water Conflicts and Challenges	Energy opportunities		Energy challenges
					Renewable source of Energy for electricity production	Oil Production (KTOE)	
				Ethiopia. Ethiopia's proposed construction of dams on the headwaters of the Blue Nile leads Egypt to repeatedly declare the vital importance of water. "The only matter that could take Egypt to war again is water" (Anwar Sadat- 1979). "The next war in our region will be over the waters of the Nile, not politics" (Boutrous Ghali- 1988)[66] Egypt sends an unsuccessful			energy, and in 1990 Egypt began mining coal in Sinai. Solar energy, an abundant resource, has not been tapped because technology is not sufficiently developed to enable it to economically compete with other energy sources.[65]

(Contd.)

(Contd. Annex-1)

African State	Population (million) /GDP Billion $	River/ Basin	Annual Renewable Water[19] Resources[20] (km^3/yr)	Water Conflicts and Challenges	Energy opportunities		Energy challenges
					Renewable source of Energy for electricity production	Oil Production (KTOE)	
				military expedition into disputed territory amidst pending negotiations over the Nile waters, Sudanese general elections, and an Egyptian vote on Sudan-Egypt unification; Nile Water Treaty signed when pro-Egyptian government elected in Sudan.[67]			

(Contd.)

(Contd. Annex-1)

African State	*Population (million) /GDP Billion $*	*River/ Basin*	*Annual Renewable Water[19] Resources[20] (km^3/yr)*	*Water Conflicts and Challenges*	*Energy opportunities*		*Energy challenges*
					Renewable source of Energy for electricity production	*Oil Production (KTOE)*	
Equatorial Guinea	0.551		26				Equatorial Guinea's natural gas reserves, as well as proven oil reserves. Renewable Energy potential is largely undeveloped, Equatorial Guinea is estimated to have 11,000 MW of hydropower potential, of which 50% is deemed economically recoverable.[69]

(Contd.)

(Contd. Annex-1)

African State	Population (million) /GDP Billion $	River/ Basin	Annual Renewable Water[19] Resources[20] (km³/yr)	Water Conflicts and Challenges	Energy opportunities		Energy challenges
					Renewable source of Energy for electricity production	Oil Production (KTOE)	
Eritrea	4.69/0.75	Gash, Nile,	6.3		Combustible renewable and waste	0	Eritrea does not produce any oil, coal or natural gas, making it dependent on petroleum product imports. The grid is supplied only by electricity from fossil fuels, though due to a lack of grid connections residential energy use consists mainly of traditional biomass.[70] Electricity is only available in Eritrea's larger cities and towns, leaving about 80% of the Eritrean population without access to

(Contd.)

(Contd. Annex-1)

African State	*Population (million) /GDP Billion $*	*River/ Basin*	*Annual Renewable Water*[19] *Resources*[20] *(km³/yr)*	*Water Conflicts and Challenges*	*Energy opportunities*		*Energy challenges*
					Renewable source of Energy for electricity production	*Oil Production (KTOE)*	
							electricity. Photovoltaic (PV) electricity generation is being used in special applications throughout the country, such as twenty-six rural health centers are each supplied with 2-kilowatt (kW) solar photovoltaic power systems for refrigeration, lighting, operating theaters, fans, and laboratory equipment. Additionally the majority of the 140 rural clinics are equipped with solar powered vaccine refrigerators.

(Contd.)

(Contd. Annex-1)

African State	Population (million) /GDP Billion $	River/ Basin	Annual Renewable Water[19] Resources[20] (km^3/yr)	Water Conflicts and Challenges	Energy opportunities		Energy challenges
					Renewable source of Energy for electricity production	Oil Production (KTOE)	
							Approximately 3% (about 60 villages) of Eritrea's villages have been supplied with PV systems (0.8 to 1.2kW) to power water pumps to supply drinking water. Each system serves a minimum of 300 households. Over 70 rural schools (out of 700) have been provided with PV systems for lighting and power. The good wind regimes found in many places in Eritrea makes it possible to save at least a part of the diesel costs by means of feeding wind electricity in the grid.[71]

(Contd.)

(Contd. Annex-1)

African State	Population (million) /GDP Billion $	River/ Basin	Annual Renewable Water[19] Resources[20] (km³/yr)	Water Conflicts and Challenges	Energy opportunities		Energy challenges
					Renewable source of Energy for electricity production	Oil Production (KTOE)	
Ethiopia	77.15/11.30	Nile, Gash, Juba-Shibeli,	110.0	Dam catchment in Ethiopia has increased rapid siltation, threatening municipal drinking water supplies in the country; Water pumping plants and pipelines in the border town of Adi Quala were destroyed during the civil war between Eritrea and Ethiopia.[72]	Hydro; Geothermal, Solar etc	0	Geothermal plant installed in 1999, is not fully operational for lack of operational experience within the community. Inspite of 60 per cent of the water in the Nile River is being supplied by the Blue Nile, which originates in the highlands of Ethiopia the country suffers from serious shortages of electricity due to slow development of hydro.

(Contd.)

(Contd. Annex-1)

African State	Population (million) /GDP Billion $	River/ Basin	Annual Renewable Water[19] Resources[20] (km^3/yr)	Water Conflicts and Challenges	Energy opportunities		Energy challenges
					Renewable source of Energy for electricity production	Oil Production (KTOE)	
				One man stabbed to death during fight over clean water during famine in Ethiopia.[73] At least 12 people died and over 20 were wounded in clashes over competition for water and pasture in the Somali border region.[74]			

(Contd.)

(Contd. Annex-1)

African State	Population (million) /GDP Billion $	River/ Basin	Annual Renewable Water[19] Resources[20] (km^3/yr)	Water Conflicts and Challenges	Energy opportunities		Energy challenges
					Renewable source of Energy for electricity production	Oil Production (KTOE)	
Gabon	1.31/5.59	Zaire,	164.0		Hydro	10929	Gabon is a large exporter of crude oil and also has natural gas resources, which are used for electricity. Only 47.9% of its population is connected with the grid. Inspite of this, Gabon still uses traditional biomass as its residential energy leading to deforestation. No new renewable energy sources like geothermal or solar are being exploited in Gabon mostly due to resource limitation. Gabon has approximately 6,000 MW of undeveloped hydroelectric potential.[75]

(Contd.)

(Contd. Annex-1)

African State	Population (million) /GDP Billion $	River/ Basin	Annual Renewable Water[19] Resources[20] (km^3/yr)	Water Conflicts and Challenges	Energy opportunities		Energy challenges
					Renewable source of Energy for electricity production	Oil Production (KTOE)	
Gambia	1.69/	Gambia	8.0				Gambia faces 40% power loss due to bad transmission lines and power theft. Due to this, fuel wood has remained its dominant energy resource, accounting for about 81% of total energy consumed. Petroleum products, including liquefied petroleum gas (LPG), accounts for about 17%; and renewable energy fewer than 1%. Use of windmills and solar PV for powering water pumps is also encouraged and is increasing throughout the country. The feasibility of using hydro-power is being examined within a sub-regional framework, in collaboration with Senegal, Guinea Bissau and Guinea Conakry.[76]

(Contd.)

(Contd. Annex-1)

African State	Population (million) /GDP Billion $	River/ Basin	Annual Renewable Water[19] Resources[20] (km^3/yr)	Water Conflicts and Challenges	Energy opportunities		Energy challenges
					Renewable source of Energy for electricity production	Oil Production (KTOE)	
Ghana	23.01/ 6.75	Volta	53.2	Declining rainfall has led to growing fights between animal herders and farmers with competing needs. In August, 2007 people were forced to flee their homes by fighting in Zounweogo province.[77]	Hydro	0	48.5% of people in Ghana are connected to grid; primary source of energy is use of traditional biomass that has lead to deforestation. Solar cookers, PV cells, small wind turbines for water pumping, and biomass gasifier for electricity generation is being used at pilot scale.[78]

(Contd.)

(Contd. Annex-1)

African State	*Population (million) /GDP Billion $*	*River/ Basin*	*Annual Renewable Water[19] Resources[20] (km³/yr)*	*Water Conflicts and Challenges*	*Energy opportunities*		*Energy challenges*
					Renewable source of Energy for electricity production	*Oil Production (KTOE)*	
Guinea Bissau	1.47/	Corubal, Geba,	31.0				Guinea-Bissau has no oil production since 1998, and is wholly dependent on refined petroleum products imported from neighbouring countries.

(Contd.)

(Contd. Annex-1)

African State	Population (million) /GDP Billion $	River/ Basin	Annual Renewable Water[19] Resources[20] (km^3/yr)	Water Conflicts and Challenges	Energy opportunities		Energy challenges
					Renewable source of Energy for electricity production	Oil Production (KTOE)	
							No natural gas or coals are consumed in Guinea-Bissau, nor does renewable energy play any part in the country's energy mix. 66.7% people use traditional fuel and 0% are currently connected to electricity grid. Installed electricity capacity in the country came from fossil fuel. Guinea-Bissau has great potential resources for hydroelectric power development for the domestic market and even for export. The potential production capacity of the Corubal and Geba rivers alone exceeds the country's estimated future needs, although due to resource limitation it hasn't been accomplished.[79]

(Contd.)

(Contd. Annex-1)

African State	*Population (million) /GDP Billion $*	*River/ Basin*	*Annual Renewable Water[19] Resources[20] (km³/yr)*	*Water Conflicts and Challenges*	*Energy opportunities*		*Energy challenges*
					Renewable source of Energy for electricity production	*Oil Production (KTOE)*	
Guinea		Corubal, Gambia, Geba, Niger, Senegal,	226.0				Guinea does not have any proven domestic reserves of coal, natural gas or oil, and is forced to import all of its fossil fuels. Guinea is the source of several major West African rivers (including the Gambia and Niger Rivers) and has a hydroelectric potential (technically feasible) estimated at 19,400 Giga watt hours per year (Gwh/yr).

(Contd.)

(Contd. Annex-1)

African State	*Population (million) /GDP Billion $*	*River/ Basin*	*Annual Renewable Water[19] Resources[20] (km³/yr)*	*Water Conflicts and Challenges*	*Energy opportunities*		*Energy challenges*
					Renewable source of Energy for electricity production	*Oil Production (KTOE)*	
Ivory Coast							
Kenya	36.55/ 16.02	Juba-Shibeli, Nile,	30.2	Police were sent to the northwestern	Hydro, Geothermal,		Kenya is highly dependent on

(Contd.)

(Contd. Annex-1)

African State	Population (million) /GDP Billion $	River/ Basin	Annual Renewable Water[19] Resources[20] (km³/yr)	Water Conflicts and Challenges	Energy opportunities		Energy challenges
					Renewable source of Energy for electricity production	Oil Production (KTOE)	
				part of Kenya to control a major violent dispute between Kikuyu and Masai groups over water. More than 20 people were killed in fighting in January 2005. By July, the death toll exceeded 90, principally in the rural center of Turbi. The tensions arose over grazing and water. Maasai herdsmen accused a local Kikuyu politician of diverting a river to irrigate his farm, depriving downstream livestock. Fighting	Solar etc		hydroelectricity, which is responsible for over 75% of all electrical output. Kenya currently does not produce crude oil. Energy consumption is dominated by wood fuel and other non-sustainable sources. Kenya has been an active market for solar home systems for almost a decade. Geothermal

(Contd.)

(Contd. Annex-1)

African State	Population (million) /GDP Billion $	River/ Basin	Annual Renewable Water[19] Resources[20] (km³/yr)	Water Conflicts and Challenges	Energy opportunities		Energy challenges
					Renewable source of Energy for electricity production	Oil Production (KTOE)	
				displaced more than 2000 villagers and reflects tensions between nomadic and settled communities.[80] A clash between villagers and thirsty monkeys left eight apes dead and ten villagers wounded. The duel started after water tankers brought water to a drought-stricken area and monkeys desperate for water attacked the villagers.[81]			electricity generation in Kenya has increased by 84 MW since 2000. 62 per cent hydropower potential has been developed in Kenya and yet only 9.1% people are connected to grid.[82]

(Contd.)

(Contd. Annex-1)

African State	*Population (million) /GDP Billion $*	*River/ Basin*	*Annual Renewable Water[19] Resources[20] (km³/yr)*	*Water Conflicts and Challenges*	*Energy opportunities*		*Energy challenges*
					Renewable source of Energy for electricity production	*Oil Production (KTOE)*	
Lesotho	2.13	Orange	5.2	The Orange basin embraces four countries but the Lesotho Highland Water Project involves only South Africa and Namibia causing a potential for conflict w/ Botswana; A coup by Lesotho's defense forces, with support from South Africa, lead to immediate agreement with South Africa for water from the Highlands of Lesotho, after 30 previous years of Unsuccessful negotiations. There is disagreement over the degree to which water was a motivating factor for either party.[83]			Lesotho has no known oil or gas reserves. Coal is the major energy source consumed in Lesotho supplying 87% of its commercial energy needs with oil only supplying 6.5%. Only 2% of the households in the country are electrified. This is a result of larg connection costs charged by utilities. There is considerable hydroelectric potential in Lesotho.[86]

(Contd.)

(Contd. Annex-1)

African State	Population (million) /GDP Billion $	River/ Basin	Annual Renewable Water[19] Resources[20] (km^3/yr)	Water Conflicts and Challenges	Energy opportunities		Energy challenges
					Renewable source of Energy for electricity production	Oil Production (KTOE)	
				South Africa supportscoup in Lesotho over support for ANC and anti-apartheid, and water. New government in Lesotho then quickly signs Lesotho Highlands Water Agreement.[84] Water infrastructure, including dams and the major Cunene-Cuvelai pipeline, was targeted during the conflicts in Namibia and Angola in the 1980s.[85]			

(Contd.)

(Contd. Annex-1)

African State	Population (million) /GDP Billion $	River/ Basin	Annual Renewable Water[19] Resources[20] (km^3/yr)	Water Conflicts and Challenges	Energy opportunities		Energy challenges
					Renewable source of Energy for electricity production	Oil Production (KTOE)	
Liberia	3.20	Mano-Morro,	232.0				Liberia has no domestic petroleum resources. Wood accounts for

(Contd.)

(Contd. Annex-1)

African State	Population (million) /GDP Billion $	River/ Basin	Annual Renewable Water[19] Resources[20] (km^3/yr)	Water Conflicts and Challenges	Energy opportunities		Energy challenges
					Renewable source of Energy for electricity production	Oil Production (KTOE)	
Liberia	3.20	Mano-Morro,	232.0				Liberia has no domestic petroleum resources. Wood accounts for about 90% of the total energy requirement. The ongoing civil unrest has caused severe fuel distribution problems and shortages. All electricity comes from fossil fuels because hydropower generators were destroyed during the civil war. Wood accounts for about 90% of the total energy requirement, a reason for deforestation.[87]

(Contd.)

(Contd. Annex-1)

African State	Population (million) /GDP Billion $	River/ Basin	Annual Renewable Water[19] Resources[20] (km^3/yr)	Water Conflicts and Challenges	Energy opportunities		Energy challenges
					Renewable source of Energy for electricity production	Oil Production (KTOE)	
Libya	6.04 / 46.46	Lake Chad	0.6		Combustible renewables and waste	85257	Libya has the largest proven oil reserves in Africa and relies heavily on it and natural gas to satisfy its energy consumption demand. Libya remains "highly unexplored" for its potential in other energy options as well such as solar, PV, wind and hydro.[58] Libya faces a severe fresh water shortage. To meet the rising demand it is anticipated to utilise desalination as an essential source of fresh water, using low cost solar conversion system, feasible for installation over very large areas with very low maintenance requirements.[59]

(Contd.)

(Contd. Annex-1)

African State	Population (million) /GDP Billion $	River/ Basin	Annual Renewable Water[19] Resources[20] (km^3/yr)	Water Conflicts and Challenges	Energy opportunities		Energy challenges
					Renewable source of Energy for electricity production	Oil Production (KTOE)	
Madgascar	19.45		337.0				Madagascar imports both crude oil as well as petroleum products to meet domestic demand. No natural gas or coal is used. Madagascar is currently suffering from a series of blackouts, following the depletion of fuel stocks. Only 0.2 per cent of the technically feasible potential of 180, 000 GWh/year hydroelectricity has been economically generated till date.[90]

(Contd.)

(Contd. Annex-1)

African State	*Population (million) /GDP Billion $*	*River/ Basin*	*Annual Renewable Water[19] Resources[20] (km³/yr)*	*Water Conflicts and Challenges*	*Energy opportunities*		*Energy challenges*
					Renewable source of Energy for electricity production	*Oil Production (KTOE)*	
Madeira (Portugal)							
Malawi	13.60	Zaire, Rovuma, Zambezi,	17.3		Hydro		86.6% of Malawi's population rely on traditional biomass as its means for energy and only 5.8% of households are currently connected to electricity.[91]

(Contd.)

(Contd. Annex-1)

African State	Population (million) /GDP Billion $	River/ Basin	Annual Renewable Water[19] Resources[20] (km^3/yr)	Water Conflicts and Challenges	Energy opportunities		Energy challenges
					Renewable source of Energy for electricity production	Oil Production (KTOE)	
Mali	11.96	Niger, Senegal, Volta,	100.0				90% of energy consumed in Mali, comes from the unsustainable use of fuel-wood. Biomass producing surface has been disappearing, leading to soil erosion and desertification, and making this the predominant environmental issue linked to energy consumption.

(Contd.)

(Contd. Annex-1)

African State	Population (million) /GDP Billion $	River/ Basin	Annual Renewable Water[19] Resources[20] (km³/yr)	Water Conflicts and Challenges	Energy opportunities		Energy challenges
					Renewable source of Energy for electricity production	Oil Production (KTOE)	
							Current efforts in the country focus on promoting promising PV applications whilst continuing development of other technologies such as solar dryers, micro hydro, wind power, small scale gasifier and biogas digesters. Hydro constitutes a large part of total produced electricity, although only 10% of the population has access to electricity.[92]

(Contd.)

(Contd. Annex-1)

African State	Population (million) /GDP Billion$	River/ Basin	Annual Renewable Water[12] Resources[10] (km³/yr)	Water Conflicts and Challenges	Energy opportunities		Energy challenges
					Renewable source of Energy for electricity production	Oil Production (KTOE)	
Mauritania	3.27	Senegal,	11.4				Mauritania imports 24,000 bbl/d of oil; and consumes significant amount of "non-commercial" (wood, biomass) energy. Mauritania gains a portion of its power from dams built on the Senegal River in a joint venture with Senegal and Mali.[93]

(Contd.)

(Contd. Annex-1)

African State	Population (million) /GDP Billion$	River/ Basin	Annual Renewable Water[19] Resources[20] (km^3/yr)	Water Conflicts and Challenges	Energy opportunities		Energy challenges
					Renewable source of Energy for electricity production	Oil Production (KTOE)	
Mauritius	1.25		2.2				Mauritius is heavily dependent upon fossil fuels for power generation, transport, industry and cooking purposes. Besides fossil fuels, bagasse is the primary energy source for domestic energy use. (Mauritius is the only country in Africa to boast full electrification, largely thanks to its early start and political commitment to electrification.) Development of hydroelectric plants has slowed due to the scarcity of water and the high installation costs.[94]

(Contd.)

(Contd. Annex-1)

African State	Population (million) /GDP Billion $	River/ Basin	Annual Renewable Water[19] Resources[20] (km^3/yr)	Water Conflicts and Challenges	Energy opportunities		Energy challenges
					Renewable source of Energy for electricity production	Oil Production (KTOE)	
Mayotte[30] (France)							
Melilla (Spain) (Autonomous City of Melilla)							
Morocco	30.50/43.88		29.0		Hydro; Geothermal, Solar etc	7	Morocco is dependent on imports to satisfy about 90% of its energy needs; imports of crude oil and petroleum products cost about $1 billion per year. Electricity production is primarily from fossil fuels accounting for 91.3% and 8.7% from hydropower. Traditional biomass is also used as an energy option.

(Contd.)

(Contd. Annex-1)

African State	*Population (million) /GDP Billion $*	*River/ Basin*	*Annual Renewable Water[19] Resources[20] (km³/yr)*	*Water Conflicts and Challenges*	*Energy opportunities*		*Energy challenges*
					Renewable source of Energy for electricity production	*Oil Production (KTOE)*	
							Morocco is promoting integration of wind farms as well as solar energy by 2012, to reduce dependency on oil imports and mitigate risks related to market uncertainties. Morocco has also initiated a challenging initiative of developing energy through desalination of seawater by wind energy and are promoting investments in this area.[95]

(Contd.)

(Contd. Annex-1)

African State	*Population (million) /GDP Billion $*	*River/ Basin*	*Annual Renewable Water*[19] *Resources*[20] *(km^3/yr)*	*Water Conflicts and Challenges*	*Energy opportunities*		*Energy challenges*
					Renewable source of Energy for electricity production	*Oil Production (KTOE)*	
Mozambique	20.91	Incomati, Limpopo, Maputo, Rovuma, Umbeluzi, Zambezi,	216.0	South Africa, Botswana and Zimbabwe have plans to extract water from its Limpopo tributaries. Mozambique has a minor share and an imbalance exists in this regard creating a potential for conflict; Most of the water resources of the Save river lie in Zimbabwe. As in the case of the Limpopo, this leaves Mozambique with an imbalance;	Hydro	0	With no present domestic sources of energy besides biomass, hydropower and a little coal, Mozambique is forced to import its fossil energy, mostly petroleum. Majority of country's electrical output is through hydropower. Cahora Bass hydroelectric facility is located on the Zambezi which provides electricity domestically, as well as to Zimbabwe and South Africa.[97]

(Contd.)

(Contd. Annex-1)

African State	Population (million) /GDP Billion $	River/ Basin	Annual Renewable Water[19] Resources[20] (km^3/yr)	Water Conflicts and Challenges	Energy opportunities		Energy challenges
					Renewable source of Energy for electricity production	Oil Production (KTOE)	
				Mozambique is denied most of the water of the Save and Limpopo. The Rovuma could be an important resource for development in this country; Dams and water transfers in the South African area of the Incomati river basin reduced freshwater flows and increased salt levels in Mozambique's Incomati estuary leading to disappearance of salt-intolerant plants and animals important to the local livelihood.			

(Contd.)

(Contd. Annex-1)

African State	*Population (million) /GDP Billion $*	*River/ Basin*	*Annual Renewable Water[19] Resources[20] (km³/yr)*	*Water Conflicts and Challenges*	*Energy opportunities*		*Energy challenges*
					Renewable source of Energy for electricity production	*Oil Production (KTOE)*	
				Regular destruction of power lines from Cahora Bassa Dam during fight for independence in the region. Dam targeted by RENAMO[96] (the Mozambican National Resistance).			
Namibia	2.06/4.46	Cunene, Okavango Makgadikgadi, Orange, Zambezi,	45.5	conflict w/ Botswana over use of pipe water from Okavango Makgadikgadi to supply its capital city with industrial	Hydro	0	Namibia does not produce any coal, oil or natural gas, and is forced to import all of its petroleum, which is mainly used for transportation and

(Contd.)

(Contd. Annex-1)

African State	*Population (million) /GDP Billion $*	*River/ Basin*	*Annual Renewable Water[19] Resources[20] (km³/yr)*	*Water Conflicts and Challenges*	*Energy opportunities*		*Energy challenges*
					Renewable source of Energy for electricity production	*Oil Production (KTOE)*	
				and drinking water[98]; potential conflict w/ Botswana for using water from Orange basin[99];			industry. Residential energy use consists primarily of traditional biomass. Namibian Govt. is experiencing severe shortage of power production due to malfunctioning of its Ruacana hydropower plant. Due to this Namibia has an agreement with South Africa to supply electricity which has also been unable to fulfill this demand due to its declining excess electricity capacity forcing Namibia to prepare itself with reduced power supply and investing in new energy sources such as solar, wind, and gas fired power plant. Currently only 34.7% households are connected to grid in Namibia.[100]

(Contd.)

(Contd. Annex-1)

African State	*Population (million) /GDP Billion $*	*River/ Basin*	*Annual Renewable Water[10] Resources[10] (km/yr)*	*Water Conflicts and Challenges*	*Energy opportunities*		*Energy challenges*
					Renewable source of Energy for electricity production	*Oil Production (KTOE)*	
Niger	12.90/	Lake Chad, Niger,	33.7				Niger currently relies heavily on traditional fuel for households – about 77.4% households use fuel wood, charcoal, bagasse (sugar cane waste) and animal and vegetable wastes as a mean for energy.

(Contd.)

(Contd. Annex-1)

African State	Population (million) /GDP Billion $	River/ Basin	Annual Renewable Water[19] Resources[20] (km³/yr)	Water Conflicts and Challenges	Energy opportunities		Energy challenges
					Renewable source of Energy for electricity production	Oil Production (KTOE)	
Nigeria	144.72/ 63.53	Lake Chad, Niger,	286.2	In the 1960s four countries shared Lake Chad: Niger,	Hydro	132210	Nigeria's natural gas reserves are the 9th largest in the world, but

(Contd.)

(Contd. Annex-1)

African State	Population (million) /GDP Billion $	River/ Basin	Annual Renewable Water[19] Resources[20] (km^3/yr)	Water Conflicts and Challenges	Energy opportunities		Energy challenges
					Renewable source of Energy for electricity production	Oil Production (KTOE)	
				Nigeria, Chad, and Cameroon. By 2001 it was shared by only two of those, Chad and Cameroon. The 1980s Nigerian fishermen began migrating into Cameroon, following the retreating waters, and in the mid 1990s more than thirtyLake Chad Basin villages founded by Nigerians were counted in Cameroon.	•		due to a lack of infrastructure, Nigeria flares 75% of the natural gas it produces. Traditional biomass (largely wood fuels) accounts for the largest share of total energy consumption (51%) followed by petroleum products (41%), natural gas (5.2%), and electricity (3.1%). Total installed electricity capacity is estimated at 5.9 GW with about 40% of the population

(Contd.)

(Contd. Annex-1)

African State	Population (million) /GDP Billion $	River/ Basin	Annual Renewable Water[19] Resources[20] (km^3/yr)	Water Conflicts and Challenges	Energy opportunities		Energy challenges
					Renewable source of Energy for electricity production	Oil Production (KTOE)	
				Tension arose between Nigeria and Cameroon when the former established state control and public services in these Nigerian-populated villages, including military and police forces, schools, and health centers.[101] A protest over the price of water in Nyanya, Abuja, Nigeria resulted in violence, including the beating of water vendors.[102]			having access to electricity (82% in urban areas and 10% in rural areas). A modest estimate of the technical potential of solar energy in Nigeria with 5% device efficiency is put at 15 x 1014 KJ of useful energy annually. This translates to about 258.62 million barrels of oil equivalent or 4.2 x 105 GWh of electricity production annually. The small hydro potential for Nigeria has been assessed at 824 MW, out which only 4% have been exploited.[103]

(Contd.)

(Contd. Annex-1)

African State	Population (million) /GDP Billion $	River/ Basin	Annual Renewable Water[19] Resources[20] (km^3/yr)	Water Conflicts and Challenges	Energy opportunities		Energy challenges
					Renewable source of Energy for electricity production	Oil Production (KTOE)	
Reunion (France)			5.0				

(Contd.)

(Contd. Annex-1)

African State	Population (million) /GDP Billion $	River/ Basin	Annual Renewable Water[19] Resources[20] (km^3/yr)	Water Conflicts and Challenges	Energy opportunities		Energy challenges
					Renewable source of Energy for electricity production	Oil Production (KTOE)	
Rwanda	9.91	Zaire, Nile,	5.2				Over 90% of Rwandans today depend on biomass for their energy. Due to the ongoing political and economic problems in Rwanda and other countries bordering the lake Kivu, there has not been a full utilisation of the lake's potential. Only 4 or 5 per cent of Rwanda's population have access to the national power grid.

(Contd.)

(Contd. Annex-1)

African State	Population (million) /GDP Billion $	River/ Basin	Annual Renewable Water[10] Resources[10] (km³/yr)	Water Conflicts and Challenges	Energy opportunities		Energy challenges
					Renewable source of Energy for electricity production	Oil Production (KTOE)	
							Rwanda has a total hydroelectric potential of approximately 100 MW, nearly a third of which has been developed at four small hydroelectric stations, and a number of independent micro-hydroelectric stations.[104]
Saint Helena							

(Contd.)

(Contd. Annex-1)

African State	*Population (million) /GDP Billion $*	*River/ Basin*	*Annual Renewable Water*[10] *Resources*[10] *(km³/yr)*	*Water Conflicts and Challenges*	*Energy opportunities*		*Energy challenges*
					Renewable source of Energy for electricity production	*Oil Production (KTOE)*	
São Tomé and Principe	0.199						Electricity is generated by thermal and hydropower, 45% and 55% respectively.

(Contd.)

(Contd. Annex-1)

African State	*Population (million) /GDP Billion $*	*River/ Basin*	*Annual Renewable Water[19] Resources[20] (km^3/yr)*	*Water Conflicts and Challenges*	*Energy opportunities*		*Energy challenges*
					Renewable source of Energy for electricity production	*Oil Production (KTOE)*	
							Approximately just over half the population of Sao Tome and Principe has access to electricity; many households are without electricity and have to rely on candlelight and kerosene lighting, as well as on biomass (firewood and charcoal) for cooking purposes.[105]
Senegal	12.07/ 5.65	Gambia, Geba, Senegal,	39.4		Hydro	0	Electricity in Senegal is generated through hydroelectric resources. Natural gas has the potential to take a more significant role as fields in Senegal are developed. All crude oil

(Contd.)

(Contd. Annex-1)

African State	*Population (million) /GDP Billion $*	*River/ Basin*	*Annual Renewable Water[19] Resources[20] (km³/yr)*	*Water Conflicts and Challenges*	*Energy opportunities*		*Energy challenges*
					Renewable source of Energy for electricity production	*Oil Production (KTOE)*	
							is imported and used in the industrial sector and electricity generation. The solar potential is partly being exploited in Senegal by introduction of PV- technology and a number of projects have been directed at using wind power for rural electrification which provides a fair amount of potential since Senegal is on the coastline. Currently only 31.4% of households in Senegal are connected to electricity grid.[108]

(Contd.)

(Contd. Annex-1)

African State	Population (million) /GDP Billion $	River/ Basin	Annual Renewable Water[19] Resources[20] (km^3/yr)	Water Conflicts and Challenges	Energy opportunities		Energy challenges
					Renewable source of Energy for electricity production	Oil Production (KTOE)	
Seychelles	0.082						The Seychelles does not produce any oil, natural gas or coals, and relies heavily on imported petroleum products to cover domestic demand. All electricity produced comes from thermal plants. In order to reduce the high expenses from oil imports the government has stimulated the use of solar water heaters. Electricity is distributed by a central government agency which also distributes water and manages sewage system in Seychelles.[107]

(Contd.)

(Contd. Annex-1)

African State	Population (million) /GDP Billion $	River/ Basin	Annual Renewable Water[19] Resources[20] (km^3/yr)	Water Conflicts and Challenges	Energy opportunities		Energy challenges
					Renewable source of Energy for electricity production	Oil Production (KTOE)	
Sierra Leone	6.14	Mano-Morro, Niger,	160.0				Approximately 80% of Sierra Leone's primary energy is derived from traditional biomass while the remaining 20% is supplied by crude oil and its by-products, namely kerosene, cooking gas and other petroleum products. Only 5% of the total population has access to electricity. 92% of households use traditional biomass as an energy option for cooking and electricity. Sierra Leone's Bumbuna hydroelectric project was destroyed during civil war and reconstructing is being impacted due to resource limitation

(Contd.)

(Contd. Annex-1)

African State	Population (million) /GDP Billion $	River/ Basin	Annual Renewable Water[19] Resources[20] (km³/yr)	Water Conflicts and Challenges	Energy opportunities		Energy challenges
					Renewable source of Energy for electricity production	Oil Production (KTOE)	
Somalia	9.12	Juba-Shibeli,	15.7	At least 250 people killed and many more injured in clashes over water wells and pastoral lands. Villagers call it the "War of the Well" and describe well warlords, well widows, and well warriors." A three-year drought has led to extensive violence over limited water resources, worsened by the lack of effective government and central planning.[108] Creation of boundaries in 1948 leaves Somali nomads under Ethiopian rule; border skirmishes occur over disputed territory in Ogaden desert where critical water and oil resources are located; cease-fire is negotiated only after			Fuel wood and agricultural residues account for 87 percent of all energy consumed, while petroleum accounted for 11.8 percent in the form of electricity. Electrical infrastructure has been damaged and destroyed, and the ongoing strife has hindered the development of new electric resources. A planned hydroelectric facility on the Juba River has been delayed due to the continued fighting. Plans for wind generation had been proposed, but were derailed following the ouster of the Barre regime.[110]

(Contd.)

(Contd. Annex-1)

African State	Population (million) /GDP Billion $	River/ Basin	Annual Renewable Water[19] Resources[20] (km^3/yr)	Water Conflicts and Challenges	Energy opportunities		Energy challenges
					Renewable source of Energy for electricity production	Oil Production (KTOE)	
				several hundred are killed[109].			
South Africa	47.39/ 168.81	Incomati, Limpopo, Maputo, Orange, Umbeluzi,	50.0	Poor delivery of water and sanitation services in Phumelela Township led to several months of protests, including some severe injuries and property damage. No one was killed during the protests, but a few people seriously injured, and municipal property was extensively damaged.[111]	Hydro, nuclear, Geothermal, Solar etc.	1500	Highest percentage of households connected to electricity i.e. 67.1% South Africa represents a wide range of investment in energy options. South Africa relies heavily on coal, and gas, as primary source of energy options along with investment in renewable energy such as hydro, geothermal and solar.[115]

(Contd.)

(Contd. Annex-1)

African State	Population (million) /GDP Billion $	River/ Basin	Annual Renewable Water[19] Resources[20] (km^3/yr)	Water Conflicts and Challenges	Energy opportunities		Energy challenges
					Renewable source of Energy for electricity production	Oil Production (KTOE)	
				A home-made bomb was discovered at a water reservoir at Wallmansthal near Pretoria. It was thought to have been meant to sabotage water supplies to Farmers.[112] Pro-apartheid council cuts off water to the Wesselton township of 50,000 blacks following protests over miserable sanitation and living conditions.[113] Union of South African troops capture Windhoek, capital of German			

(Contd.)

(Contd. Annex-1)

African States	*Population (million) /GDP Billion $*	*River/ Basin*	*Annual Renewable Water[19] Resources[20] (km³/yr)*	*Water Conflicts and Challenges*	*Energy opportunities*		*Energy challenges*
					Renewable sources of Energy for electricity production	*Oil Production (KTOE)*	
				Southwest Africa. (May.) Retreating German troops poison wells – "a violation of the Hague convention".[114]			

(Contd.)

(Contd. Annex-1)

African State	Population (million) /GDP Billion $	River/ Basin	Annual Renewable Water[19] Resources[20] (km³/yr)	Water Conflicts and Challenges	Energy opportunities		Energy challenges
					Renewable source of Energy for electricity production	Oil Production (KTOE)	
Sudan	37.71/ 18.93	Nile, Gash, Lake Chad,	154.0	The ongoing civil war in the Sudan has included violence against water resources. In 2003, villagers from around Tina said that bombings had destroyed water wells. In Khasan Basao they alleged	Hydro	16390	Sudan is a large producer of crude oil, of which over 70% is directly exported. Sudan does not produce or import any natural gas or coal. Both residential as well as industry energy use is dominated by wood fuel.

(Contd.)

(Contd. Annex-1)

African State	Population (million) /GDP Billion $	River/ Basin	Annual Renewable Water[19] Resources[20] (km³/yr)	Water Conflicts and Challenges	Energy opportunities		Energy challenges
					Renewable source of Energy for electricity production	Oil Production (KTOE)	
				that water wells were poisoned. In 2004, wells in Darfur were Intentionally contaminated as part of a strategy of harassment against displaced populations.[116] Demonstrations in Juba, Sudan in 1978 opposing the construction of the Jonglei Canal led to the deaths of two students. Construction of the Jonglei Canal in the Sudan was forcibly suspended in 1984 following a series of attacks on the construction site.[117]			GEF has attempted to initiate implementation of solar projects to replace diesel systems. An interconnected power grid, involving hydro plants in the Upper Nile basin and the Egyptian grid could solve problems of instability and insufficient capacity supplied to Sudan.[118]

(Contd.)

(Contd. Annex-1)

African State	*Population (million) /GDP Billion $*	*River/ Basin*	*Annual Renewable Water[19] Resources[20] (km³/yr)*	*Water Conflicts and Challenges*	*Energy opportunities*		*Energy challenges*
					Renewable source of Energy for electricity production	*Oil Production (KTOE)*	
Swaziland	1.13	Incomati, Maputo, Umbeluzi,	4.5				Swaziland holds coal reserves which are being exploited for domestic consumption. As no oil reserves are present, all petroleum products are imported from South Africa. Over 75% of the population lives in rural areas, which use wood fuel as their main source of energy for cooking and heating, using inefficient stoves. The Renewable Energy Association of Swaziland (REASWA) aims to introduce renewable energy sources to the rural population and reduce pressure on the environment, using wind, solar, hydro and biomass, but is unable to do so due to limited resources.[119]

(Contd.)

(Contd. Annex-1)

African State	*Population (million) /GDP Billion $*	*River/ Basin*	*Annual Renewable Water[19] Resources[20] (km³/yr)*	*Water Conflicts and Challenges*	*Energy opportunities*		*Energy challenges*
					Renewable source of Energy for electricity production	*Oil Production (KTOE)*	
Tanzania	39.46/ 13.20	Zaire, Nile, Rovuma, Zambezi,	91	Tanzania, is denied most of the water of the Save and Limpopo river. The Rovuma could be an important resource for development in this country.	Hydro	0	Tanzanian government has attempted to diversify the country's sources of energy, but so far with limited successes. The country's electricity supply has been erratic because of the national grid's heavy reliance on hydroelectric power, which in turn depends on rainfall. In the past year, poor rainfall contributed to several electricity shortages due to the inability of the country's Ubungo electricity plant

(Contd.)

(Contd. Annex-1)

African State	Population (million) /GDP Billion $	River/ Basin	Annual Renewable Water[19] Resources[20] (km^3/yr)	Water Conflicts and Challenges	Energy opportunities		Energy challenges
					Renewable source of Energy for electricity production	Oil Production (KTOE)	
							to supply adequate amounts of energy throughout the country. Only three quarters of the country (mainly urban areas) is connected to the national grid. Tanzania relies exclusively on imports for its oil needs. In contrast to oil, Tanzania is the only country in the Great Lakes region to have considerable domestic reserves of coal. Very little attempt has been made to utilise wind and solar energy.

(Contd.)

(Contd. Annex-1)

African State	Population (million) /GDP Billion $	River/ Basin	Annual Renewable Water[19] Resources[20] (km^3/yr)	Water Conflicts and Challenges	Energy opportunities		Energy challenges
					Renewable source of Energy for electricity production	Oil Production (KTOE)	
							Tanzania announced attractive financial terms for potential investors in developing its vast renewable energy resources. It has simplified procedures for investing in solar, wind and micro-hydro projects including a 100% depreciation allowance in the first year of operation, exemption from excise duty and sales tax and concessionary customs duty on the first import of materials used in renewable energy projects.[120]

(Contd.)

(Contd. Annex-1)

African State	Population (million) /GDP Billion $	River/ Basin	Annual Renewable Water[19] Resources[20] (km^3/yr)	Water Conflicts and Challenges	Energy opportunities		Energy challenges
					Renewable source of Energy for electricity production	Oil Production (KTOE)	
Togo	6.41/ 1.5	Volta	14.7		Hydro	0	With no proven oil or natural gas reserves, Togo is forced to import all of its petroleum, which is the only fossil energy used in Togo. Electricity is generated mainly from hydropower and petroleum. Combustibles like fuel wood are used for residential energy consumption, which is the main cause for deforestation. Togo does not utilise any type of new renewable energy source like geothermal energy or wind energy.[121]

(Contd.)

(Contd. Annex-1)

African State	Population (million) /GDP Billion $	River/ Basin	Annual Renewable Water[10] Resources[10] (km^3/yr)	Water Conflicts and Challenges	Energy opportunities		Energy challenges
					Renewable source of Energy for electricity production	Oil Production (KTOE)	
Tunisia	10.13/ 25.45		4.6		Hydro Geothermal, Solar etc	3551	

(Contd.)

(Contd. Annex-1)

African State	Population (million) /GDP Billion $	River/ Basin	Annual Renewable Water[19] Resources[20] (km³/yr)	Water Conflicts and Challenges	Energy opportunities		Energy challenges
					Renewable source of Energy for electricity production	Oil Production (KTOE)	
Uganda	30.62	Nile,	66.0				Uganda consumes (about 7,000 bbl/d) from Kenya's Mombasa refinery, which in turn imports crude oil from abroad. Much of the electricity network in Uganda is poorly maintained and power cuts are frequent. Just 3-5% of the population has regular access to electricity and many towns, especially in the North of the country do not have power supply.

(Contd.)

(Contd. Annex-1)

African State	Population (million) /GDP Billion $	River/ Basin	Annual Renewable Water[19] Resources[20] (km³/yr)	Water Conflicts and Challenges	Energy opportunities		Energy challenges
					Renewable source of Energy for electricity production	Oil Production (KTOE)	
							There are numerous potential micro hydro stations throughout the country, however, external funding agencies consider these too small or uneconomic to develop.[122]

(Contd.)

(Contd. Annex-1)

African State	*Population (million) /GDP Billion $*	*River/ Basin*	*Annual Renewable Water[10] Resources[10] (km³/yr)*	*Water Conflicts and Challenges*	*Energy opportunities*		*Energy challenges*
					Renewable source of Energy for electricity production	*Oil Production (KTOE)*	
Western Sahara (*Sahrawi Arab Democratic Republic*)	0.383						

(Contd.)

(Contd. Annex-1)

African State	*Population (million) /GDP Billion $*	*River/ Basin*	*Annual Renewable Water[19] Resources[20] (km^3/yr)*	*Water Conflicts and Challenges*	*Energy opportunities*		*Energy challenges*
					Renewable source of Energy for electricity production	*Oil Production (KTOE)*	
Zambia	11.70/ 4.34	Zaire, Zambezi,	105.2	Zambezi runs through eight member states of SADC. The Zambezi River Authority (ZRA) involves onlytwo (Zimbabwe and Zambia) of these states. Angola and Zambia did not sign the SADC Protocol on Shared Watercourse Systems; Bomb blast destroyed themain water pipeline, cutting off water for the city of Lusaka, population 3 million.[123]	Hydro	0	The largest source of energy in Zambia is traditional biomass. Residential energy use consists primarily of wood fuel. No new renewable sources like geothermal or wind are present in Zambia. Zambia has abundant hydroelectric resources and meets most of its energy needs from its own hydroelectric stations. Zambia imports all of its fossil fuels, apart from a small amount of coal.[124]

(Contd.)

Literature referenced

1. Energy Demands on Water Resources, Report to U.S. Congress on the Interdependency of Energy and Water, U.S. Department of Energy, December 2006.

2. Water Energy Relationship, In support of the 2005 Integrated Energy Policy Report, California Energy Commission.

3. Energy Down the Drain, The Hidden Costs of California's Water Supply, NRDC and Pacific Institute, August 2004.

4. The Connection: Water and Energy Security; Dr. Allan R. Hoffman, Senior Analyst, U.S. Department of Energy (DOE), served as associate and acting deputy assistant secretary for Utility Technologies in the Office of Energy Efficiency and Renewable Energy of the DOE and is an IAGS Advisor. http://www.iags.org/n0813043.htm

5. Considerations on Liquid Biofuels as a Viable Alternative to Fossil Fuels, Prepared by key staff of the National Renewable Energy Laboratory's National Bio-energy Center, Presented by Director Dan E. Arvizu.

6. Liquid biofuels as a viable alternative to fossil fuels, by Thomson Sinkala, UNIDO.

7. Sustainable Industrial Conversion and Productive Uses of Biofuels, UNIDO's Biofuel Strategy.

8. Refrigerator efficiency in Ghana: Tailoring an appliance market transformation programme design for Africa, Robert Van Buskirk, Essel Ben Haganb, Alfred Ofosu Ahenkorahc, Michael A. McNeil; Lawrence Berkeley National Laboratory, One Cyclotron Road, MS: 90R4000, Berkeley, CA 94720, USA;CSIR Institute for Industrial Research, P.O. Box LG 576, Legon, Accra, Ghana; Ghana Energy Foundation, 5 East Legon, P.O. Box CT1671, Cantonments, Accra, Ghana; Lawrence Berkeley National Laboratory, One Cyclotron Road, MS: 90R4000, Berkeley, CA 94720, USA.

9. Analysis of long-range clean energy investment scenarios for Eritrea, East Africa; Robert Van Buskirk, Lawrence Berkeley National Laboratory, Energy Analysis Department, MS 90-4000, 1 Cyclotron Road, Berkeley, CA 94720, USA.

10. Biofuels for poverty reduction, Prof. Thomson Sinkala; Chairman, Biofuels Association of Zambia.

11. Biofuels for Industrial Development and Poverty Reduction in Africa; Heinz Leuenberger Director, Energy and Cleaner Production Branch, United Nations Industrial Development Organisation (UNIDO).

12. Clean Energy Trends 2006.

13. Sustainable Bio-energy: A Framework for Decision Makers; UN Energy; April 2007.

14. Discussion paper on GBEP facilitation of collaborative bio-energy field projects, UNDP Global Bio-energy Partnership.

15. Discussion paper on a proposal for a GBEP Task Force to work towards the development of a voluntary international sustainability framework of principles for bio-energy.

16. Valuing and Charging for Water, UNESCO report on Water.

17. Luis Gomez-Echeverri, Director, a.i. Bureau for Development Policy, United Nations Development Programme, Capacity Development for Expanding Energy Access in Africa.

18. Improved Access and Pro-Poor Energy Policy, Africa-Europe Energy Forum, March 7, 2007, S. Vijay Iyer, World Bank.

19. White Paper for Regional Policy : Geared towards increasing access to energy services for rural and peri-urban populations in order to achieve the Millennium Development Goals, The drafting and publication of this White Paper have been made possible thanks to both UNDP and the French Ministry of Foreign Affairs, ECOWAS/UEMOA; ftp://ftp2.ptfm.net/ptfmc/references/ecowas_white_paper.pdf

20. Regulation and Private Participation in the Water and Sanitation Sector, by JUDITH A. REES; Global Water Partnership Technical Advisory Committee, Swedish International Development Cooperation Agency.

21. Tackling Global Water Priorities, Global Water Partnership.

22. Energy Report, ENERDATA, December 2006, Enerdata sas – 2, Avenue de Vignate – 38610 Gières – FRANCE www.enerdata.fr

23. Country Report for Regional Policy Based On "Increasing Acccess to Energy Services for Populations in Rural and Sub Urban Areas in Order to Achieve The Millennium Development Goals" Prepared by KITE, For UNDP-REPP and ECOWAS, October, 2005.

24. Energy Access in Sub-Saharan Africa – A World Bank Action Plan Programme of Action for The Least Developed Countries (2001-2010), Geneva, July 18-19, 2006, Africa Energy Unit The World Bank, Mac Cosgrove-Davies, Sr. Energy Specialist.

25. Cabraal, Anil; Davies, Mac Cosgrove and Schaeffer, Loretta, "Best Practices for Photovoltaic Household Electrification Programs: Lessons From Experiences in Selected Countries", (1996), The World Bank, Washington D.C., USA. http://www.worldbank.org/astae/pvpdf/cover.htm

26. Energy Efficiency Policies around the World: Review and Evaluation World Energy Council 2008, Promoting the sustainable supply and use of energy for the greatest benefit of all.

27. Water for Energy and Energy for Water, WISIONS for Sustainability, 2008, Wuppertal Institute for Climate, Environment and Energy, Germany.

28. Energy Services for Millennium Development Goals, UN Millennium Project, UNDP, World Bank, and ESMAP, 2005.

29. Freshwater Withdrawal by Country and Sector, 2007 Pacific Institute.

30. GEF Strategic Programme for West Africa, Program Framework Document, The GEF Trust Fund, 2008.

31. GEF Strategic Programme for West Africa (GEF-SPWA) Biodiversity and Climate Change Focal Areas; 2008.

32. Discussion paper on a proposal for a GBEP Task Force to work towards the development of a voluntary international sustainability framework of principles for bio-energy; GBEP Global Bio-energy Partnership, 2008.

33. IDA At Work; Burkina Faso: Overcoming the Odds, March 2007; http://www.worldbank.org/ida

34. Renewable Energy Policy and Wind Energy in Morocco, Amal Haddouche, General Director of CDER (Morocco). Presentation delivered in Berlin, October 18-19, 2006.

35. Briefing Statement "The Global Water Crisis" Olav Kjorven Director, Energy and Environment Group Bureau for Development Policy United Nations Development Programme (UNDP) , House Committee on International Relations United States Congress 29 June 2005.

36. Renewable Energy Technologies and Poverty Alleviation: Overcoming Barriers and Unlocking Potentials, Global Network On Energy For Sustainable Development; 2002.

37. Politics of Water: Conflict and Cooperation in Africa: Dr. Marcel Kitissou.

38. Prospects of Renewable Energy in Libya, Ibrahim M. Saleh, Faculty of Engineering, Al-Fateh University P.O. Box 13656 Tripoli, Libya.

39. Renewables 2007, Global Status Report, Renewable Energy Policy Network for the 21st Century.

40. Southwest Hydrology Newsletter, Resource for Semi-Arid Hydrology. A bimonthly trade magazine for hydrologists, water managers, and other professionals working with water issues; September 2007.

41. Sustainable Bio-energy: A Framework for Decision Makers, United Nations/UN Energy.

42. Sustainable Energy and Community Participation in Biomass-based Cogeneration in Africa Compiled by Stephen Karekezi, John Kimani, Justus Mbithi, Jared Atiang and Oscar Onguru.

43. Tackling urgent water priorities, Global Water Partnership.

44. The Role of Small and Medium-Scale Renewable Energy Technologies in Poverty Alleviation, Environmental Sustainability and Economic Development in Ethiopia by Workeneh Gashie.

45. First Joint Annual Meetings of the AU Conference of Ministers of Economy and Finance and ECA Conference of African Ministers of Finance, Planning and Economic Development, Addis Ababa, Ethiopia; Twenty-seventh meeting of the Committee of Experts 26 – 29 March 2008.

46. UNESCO Water : A Shared Responsibility report.

47. Water and Energy Nexus, Michael R. Gabaldon, Director, Technical Service Center, U.S. Department of the Interior, Bureau of Reclamation.

48. Wolf, A., S. Yoffe, M. Giordano. "International Waters: Identifying Basins at Risk." Water Policy. Vol. 5 #1, 2003. pp. 31-62.

49. The Water-Energy Nexus: Opportunities for Integrated Environmental Management; USAID Global Environment Center, July 2001.

50. Volta River Authority, Ministry of Energy, Govt. of Ghana, Brief profiles of Sector Institutions, December 2003.

51. Water-Energy Relationship, In support of the 2005 Integrated Energy Policy Report Matt Trask Environmental Office Systems Assessment and Facilities Siting Division California Energy Commission, June 2005.

52. Wind Power for Rural Energy Services in Eritrea, Desalinisation, Technical Advisor for Eritrea Wind

Energy Applications Project Prepared by Douglas M. Bushey Lawrence Berkeley National Laboratory UC Berkeley Energy and Resources Group, June 2006.

53. Wind Power for Rural Energy Services in Eritrea, Water Pumping, Technical Advisor for Eritrea Wind Energy Applications Project Prepared by Douglas M. Bushey Lawrence Berkeley National Laboratory UC Berkeley Energy and Resources Group, June 2006.

54. World Energy Outlook 2005: Fact Sheet What Are The Key Energy Trends In The Middle East And North Africa?

55. Southern African Water Conflicts:Are They Inevitable Or Preventable? Peter Ashton, Green Cross International, 2000, Water for Peace in the Middle East and Southern Africa, Green Cross International, Geneva, pp. 94-98.

56. The Global Water Partnership Addressing Challenges of Globalization: An Independent Evaluation of the World Bank's Approach to Global Programs Case Study, Saeed Rana and Lauren Kelly.

57. Renewable Water Resources, typically comprised of both renewable surface water and groundwater supplies, including surface inflows from neighbouring countries.

58. The World's Water series, 2006.

59. IEA Energy Statistics for Algeria.

60. Ministry of Energy and Mining, Government of Algeria.

61. IEA Angola Country information.

62. Government of Angola, Ministry of Energy and Water http://www.minea.gv.ao/

63. International Herald Tribune, 1999; Water Conflict Chronology, the Pacific Institute.

64. Meissner 2001, and Water Conflict Chronology, the Pacific Institute.

65. Meissner 2000, and the Water Conflict Chronology, the Pacific Institute.
66. Meissner 2001, and Water Conflict Chronology, the Pacific Institute.
67. http://www.unitedworld-usa.com/reports/angola/infrastructure.asp
68. http://www.unitedworld-usa.com/reports/angola/infrastructure.asp
69. Country Energy Report: Angola http://www.energy recipes. org/reports/countryreports.php?action=all 27 Meissner 2001.
70. http://www.gawa.nu/stories/story_solar.htm *Promotion of Solar Energy in Benin* Importance of networks and partnership; creating a replicable model for encouraging solar energy.
71. Country Energy Report: Benin, http://www.energy recipes.org/reports/genericData/Africa/061129% 20RECIPES% 20country% 20info%20Benin.pdf.
72. Hydropolitical Vulnerability and Resilience along International Waters: Africa.
73. Dr. Peter Ashton, the Division of Water, Environment & Forestry Technology, CSIR, P.O. Box 395, Pretoria 0001, South Africa.
74. http://www.sovereign-publications.com/botswana.htm.
75. http://www.sovereign-publications.com/botswana.htm.
76. Professor Richard O. Fagbenle, Republic of Botswana, Energy Affairs Division, Ministry of Minerals, Energy and Water Affairs.
77. IDA at Work: Burkina Faso: Overcoming the Odds.
78. UNOCHA, 2007.

79. Country energy information: Burkina Faso, Developing Renewable, September 2006 http://www.energy recipes.org/reports/genericData/Africa/061129% 20RECIPES% 20country%20info% 20Burkina %20 Faso.pdf.

80. Country energy information: Burkina Faso, Developing Renewable, September 2006 http://www.energyrecipes.org/reports/genericData/Africa/061129%20RECIPES% 20country% 20info% 20Burkina %20Faso.pdf.

81. http://en.wikipedia.org/wiki/Mining_in_Burundi.

82. http://www.energyrecipes.org/reports/genericData/Africa/061129%20RECIPES%20country%20info%20Burundi.pdf.

83. http://www.energyrecipes.org/reports/genericData/Africa/061129%20RECIPES%20country%20info%20Burundi.pdf.

84. http://www.cameroon-info.net/cmi_show_news.php?id=21689.

85. Cameroon: Bini à Warak Hydroelectric Power Plant Feasibility Study.

86. Cameroon: Limb Gas to Electricity Technical Assistance, Oct. 2008, USTDA.

87. Cameroon: Limb Gas to Electricity Technical Assistance, Oct. 2008, USTDA.

88. Country Energy Information: Cape Verde; Developing Renewables, September 2006.

89. Country Energy Information: Central African Republic, September 2006.

90. Country Energy Information: Central African Republic, September 2006.

91. Country Energy Report: Chad, Developing Renewables, September 2006.

92. Country Energy Report: Chad, Developing Renewables, September 2006.

93. Country Energy Report: Chad, Developing Renewables, September 2006.

94. Country Energy Report: Comoros, Developing Renewables, September 2006.

95. UNOCHA, 2007.

96. Country Energy Report: Cote d'Ivoire, Developing Renewables, September 2006.

97. Country Energy Report, Cote d'Ivoire; Developing Renewables, September 2006.

98. Chenje, 2001, Human Rights Watch 1998, and Water Conflict Chronology, the Pacific Institute.

99. Country Energy Report, DRC; Developing Renewables, September 2006.

100. Country Energy Report: Djibouti, Developing Renewables, September 2006.

101. Country Energy Report: Djibouti, Developing Renewables, September 2006.

102. http://www.iasps.org/strategic/water.htm.

103. Gleick 1991, 1994.

104. http://countrystudies.us/egypt/91.htm.

105. Wolf 1997.

106. Country Energy Report: Equatorial Guinea, Developing Renewables, September 2006.

107. Country Energy Report: Eritrea, Developing Renewables, September 2006.

108. Country Energy Report: Eritrea, Developing Renewables, September 2006.

109. ICRC 2003 and Water Conflict Chronology, the Pacific Institute.

110. Sandrasagra 2000 and Water Conflict Chronology, the Pacific Institute.

111. BBC 2006a and Conflict Chronology, the Pacific Institute.

112. Country Energy Information: Gabon, Developing Renewables, September 2006.

113. Country Energy Information: Gambia, Developing Renewables, September 2006.

114. UNOCHA, 2007.

115. Energy Country Report: Ghana, Developing Renewables, September 2006.

116. Energy Country Report: Guinea Bissau, Developing Renewables, September 2006.

117. BBC 2005a, Ryu 2005, Lane 2005; Water Conflict Chronology, The Pacific Institute.

118. BBC 2005, and Okoko 2000; Water Conflict Chronology, The Pacific Institute.

119. Country Energy Information, Kenya; Developing Renewables September 2006.

120. Mohammed 2001, and the World Water Chronology, the Pacific Institute.

121. American University 2000b.

122. Turton 2005; Pacific Institute.

123. Country Energy Information: Lesotho, Developing Renewables, September 2006.

124. Country Energy Information: Liberia, Developing Renewables, September 2006.

125. Libya Energy profile: http://www.energypublisher.com/article.asp?id=10484.

126. Renewable Energy in Libya: Status and Prospects, F. M. F. Siala and M. A. Musa, Center for Solar Energy Studies, Tripoli, Libya.

127. Country Energy Information: Madagascar, Developing Renewables, September 2006.

128. Country Energy Information: Malawi, Developing Renewables, September 2006.

129. Country Energy Information: Mali, Developing Renewables, September 2006.

130. Country Energy Information: Mauritania, Developing Renewables, September 2006.

131. Country Energy Information: Mauritius, Developing Renewables, September 2006.

132. RENEWABLE ENERGY POLICY AND WIND ENERGY IN MOROCCO, presentation by Amal Haddouche; General Director of CDER (Morocco).

133. Chenje 2001.

134. Country Energy Information: Mozambique, Developing Renewables, September 2006.

135. Prof. Aaron Wolf, Hydro political Vulnerability and Resilience along International Waters: Africa; A Study conducted by UNEP, and Oregon State University.

136. Hydro political Vulnerability and Resilience along International Waters: Africa, A study by UNEP and Oregon State University.

137. Country Energy Information: Namibia, Developing Renewables, September 2006.

138. The Stimson Institute, The Lake Chad Water Conflict Study.

139. Yakubu, 2008 and Water Conflict Chronology the Pacific Institute.

140. Country Energy Information: Nigeria, Developing Renewables, September 2006.

141. Country Energy Information: Rwanda, Developing Renewables, September 2006.

142. Country Energy Information: São Tomé and Principe, Developing Renewables, September 2006.

143. Country Energy Information: Senegal, Developing Renewables, September 2006.

144. Country Energy Information: Seychelles, Developing Renewables, September 2006.

145. BBC 2004, AP 2005, WAX 2006; and Water Conflict Chronology the Pacific Institute.

146. Country Energy Information: Seychelles, Developing Renewables, September 2006.

147. Wolf 1997.

148. CDE 2007, Water Conflict Chronology, the Pacific Institute.

149. Pretoria Dispatch, 1999 and the Water Conflict Chronology, the Pacific Institute.

150. Gleick 1993 and the Water Conflict Chronology, the Pacific Institute.

151. Country Energy Information: South Africa, Developing Renewables, September 2006.

152. Daniel 1995 and the Pacific Institute.

153. Toronto daily 2004, Reuters Foundation 2004 and Water Conflict Chronology the Pacific Institute.

154. Suliman1998; Keluel-Jang 1997.

155. Country Energy Information: Sudan, Developing Renewables, September 2006.

156. Country Energy Information: Swaziland, Developing Renewables, September 2006.

120. Country Energy Information: Tanzania, Developing Renewables, September 2006.

121. Country Energy Information: Togo, Developing Renewables, September 2006.

122. Country Energy Information: Uganda, Developing Renewables, September 2006.

123. FTGWR, 1999.

124. Country Energy Information: Zambia, Developing Renewables, September 2006.

125. Chenje 2001.

126. Country Energy Information: Zimbabwe, Developing Renewables, September 2006.

Annexure : Brief Information Regarding 274 GEF Projects in 44 Countries of Africa

Region or Country	*Project Title*	*Primary Focal Area*	*Operational Programme*	*Project Type*	*Project Status*
Africa : ANGOLA Angola, Botswana, Namibia	Environment protection and sustainable management of the Okavango River Basin	International Waters	9: Integrated Land and Water	FULL Size Projects	Project being implemented
Africa : ANGOLA	Biodiversity Strategy and Action Plan	Biodiversity	BD: Enabling Activities	IDEA	Project has been completed
Africa : ANGOLA	Energy and livelihoods solutions Angola (Elisa) – Sustainable Land Management capacity building and combating deforestation	Land Degradation	15: Sustainable Land Management	MEDIUM Size Projects	Project being implemented
Africa : BURUNDI Burundi, Congo Democratic Republic, Tanzania, United Republic of, Zambia	Pollution control and other measures to protect biodiversity in Lake Tanganyika	International Waters Mitigation	8: Waterbody	IDEA	Project has been completed

(Contd.)

(Contd. Annexure-II)

Region or Country	*Project Title*	*Primary Focal Area*	*Operational Programme*	*Project Type*	*Project Status*
Africa : BURUNDI	Add-on 3: Capacity Needs Assessment for Implementation of BSAP and CHM Support	Biodiversity	BD: Enabling Activities	ENABLING Activities	Project being implemented
Africa : BURUNDI	ADD-ON: Enabling Burundi to prepare its first national communication to the UNFCCC	Climate Change - Mitigation	CC: Enabling Activities	ENABLING Activities	Project being implemented
Africa : BURUNDI	National Self-assessment of capacities for global environment management		Capacity Enabling Activities	ENABLING Activities	Project being implemented
Africa : BURUNDI	Preparation of a National Plan of Action for Adaptation for Burundi for Climate Change	Climate Change - LDCF		ENABLING Activities	Project being implemented
Africa : BURUNDI	Burundi Second National Communications to the UNFCCC	Climate Change -	CC: Enabling Activities	ENABLING Activities	Project being implemented

(Contd.)

(Contd. Annexure-II)

Region or Country	*Project Title*	*Primary Focal Area*	*Operational Programme*	*Project Type*	*Project Status*
Africa : BENIN	Participatory Management of Natural Forests and Village Reforestation to reduce Carbon Emissions	Climate Change - Mitigation	5: Conservation and Efficiency	IDEA	Project has been completed
Africa : BENIN	Benin BD Enabling Activities	Biodiversity	BD: Enabling Activities	IDEA	Project has been completed
Africa : BENIN	Climate Change Enabling Activity (additional financing for capacity building in priority areas)	Climate Change - Mitigation	CC: Enabling Activities	ENABLING Activities	Project being implemented
Africa : BENIN	Self-Assessment of National Capacity Building Requirements for Global Environment Management in Benin	Multiple Focal Area			

(Contd.)

(Contd. Annexure-II)

Africa : BENIN	National Adaptation Programme of Action (LDCs	Climate Change - LDCF	CC: Enabling Activities	IDEA	Project has been completed
Africa : BENIN	Benin: Second national communication to the UNFCCC	Climate Change - Mitigation	CC: Enabling Activities	ENABLING Activities	Project being implemented
Africa : BURKINA FASO	Optimisation of biodiversity in game ranching systems; a pilot experiment in a semi arid area	Biodiversity	1: Arid & semi-arid ecosystems	IDEA	Project has been completed
Africa : BURKINA FASO Burkina Faso, Cameroon, Republic of, Ethiopia, Ghana, Kenya, Madagascar, Sierra Leone, South Africa, Uganda	Building non-governmental organization capacity through biodiversity survey in Africa	Biodiversity	3: Forest ecosystems	IDEA	Project has been completed

(Contd.)

(Contd. Annexure-II)

Region or Country	*Project Title*	*Primary Focal Area*	*Operational Programme*	*Project Type*	*Project Status*
Africa : BURKINA FASO	Burkina Faso BD Enabling Activities	Biodiversity	BD: Enabling Activities	IDEA	Project has been completed
Africa : BURKINA FASO	ADD-ON: Expedited Financing for (Interim) Measures for Capacity Building in Priority Areas Part II	Climate Change - Mitigation	CC: Enabling Activities	ENABLING Activities	Project being implemented
Africa : BURKINA FASO	Energy Sector Reform and Development	Climate Change - Mitigation	6: Renewables	PDF B	Project being developed
Africa : BURKINA FASO Benin, Niger	Enhancing the effectiveness and catalysing the sustainability of the W-Arly-Pendjari (WAP) protected area system	Biodiversity	1: Arid & semi-arid ecosystems	FULL Size Projects	Project being implemented

(Contd.)

(Contd. Annexure-II)

Region or Country	*Project Title*	*Primary Focal Area*	*Operational Programme*	*Project Type*	*Project Status*
Africa : BURKINA FASO	Enabling activity for the preparation of a National Implementation Plan (NIP) on POPs for Burkina Faso	POPS	14: Persistent Organic Pollutants	ENABLING Activities	Project being implemented
Africa : BURKINA FASO	National Capacity Needs Self Assessment for Global Environmental Management for Burkina Faso	Multiple Focal Area	Capacity Enabling Activities	ENABLING Activities	Project being implemented
Africa : BURKINA FASO	Expedited Funding of National Adaptation Programmes of Action by Least Developed Countries	Climate Change - LDCF	CC: Enabling Activities	ENABLING Activities	Project being implemented

(Contd.)

(Contd. Annexure-II)

Region or Country	*Project Title*	*Primary Focal Area*	*Operational Programme*	*Project Type*	*Project Status*
Africa : BURKINA FASO	Country Partnership Programme on Sustainable Land Management in Burkina Faso	Land Degradation	15: Sustainable Land Management	PDF B	Project being developed
Africa : BURKINA FASO	Burkina Faso: Second Communication National to the UNFCCC	Climate Change - Mitigation	CC: Enabling Activities	ENABLING Activities	Project being implemented
Africa : BOTSWANA	Enabling Botswana to fulfil its commitments to the UN Framework Convention on Climate Change (UNFCCC)	Climate Change - Mitigation	CC: Enabling Activities	ENABLING Activities	Project has been completed
Africa : BOTSWANA Botswana, Lesotho, Mozambique, South Africa, Swaziland, Zambia, Zimbabwe	Southern African Biodiversity Support Programme	Biodiversity	3: Forest ecosystems	FULL Size Projects	Project being implemented

(Contd.)

(Contd. Annexure-II)

Region or Country	*Project Title*	*Primary Focal Area*	*Operational Programme*	*Project Type*	*Project Status*
Africa : BOTSWANA	Botswana Biological Diversity Enabling Activity (Additional financing for assessing capacity building needs)	Biodiversity	BD: Enabling Activities	ENABLING Activities	Project being implemented
Africa : BOTSWANA Botswana, Kenya, Mali	Management of Indigenous Vegetation for the Rehabilitation of Degraded Rangelands in the Arid Zone of Africa -FULL	Biodiversity	1: Arid & semi-arid ecosystems	FULL Size Projects	Project being implemented

(Contd.)

(Contd. Annexure-II)

Region or Country	*Project Title*	*Primary Focal Area*	*Operational Programme*	*Project Type*	*Project Status*
Africa : CENTRAL AFRICAN REPUBLIC	Strengthening institutional capacities to combat deforestation land degradation in the Central African Republic	Land Degradation	15: Sustainable Land Management	MEDIUM Size Projects	Project being implemented
Africa : CHAD Cameroon, Republic of, Central African Republic, Niger, Nigeria	Reversal of land and water degradation trends in the Lake Chad Basin ecosystem	International Waters	9: Integrated Land and Water	FULL Size Projects	Project being implemented
Africa : CHAD	Add On 2 - Identification of Capacity-Building needs for the Implementation of the national BSAP	Biodiversity	BD: Enabling Activities	ENABLING Activities	Project being implemented
Africa : CHAD	Conservation and sustainable use of biodiversity in the Moyen-Chari	Biodiversity	1: Arid & semi-arid ecosystems	FULL Size Projects	Project being implemented
Africa : CHAD	Enabling Chad to prepare its first national communication in response to its commitment to the UNFCCC -ENABLING	Climate Change - Mitigation	CC: Enabling Activities	ENABLING Activities	Project being implemented

(Contd.)

(Contd. Annexure-II)

Region or Country	*Project Title*	*Primary Focal Area*	*Operational Programme*	*Project Type*	*Project Status*
Africa : CHAD, Cameroon, Republic of, Central African Republic, Chad, Niger, Nigeria	Degradation in the Lake Chad Basin ecosystem Reversing land and water	Waters International	8: Waterbody	PDF C	Developed Project being
Africa : CHAD	BRIEF NCSA Chad	Multiple Focal Area	Capacity Enabling Activities	ENABLING Activities	Project being implemented
Africa : CHAD	Preparation of a National Programme of Action for Adaptation to Climate Change	Climate Change - LDCF	CC: Enabling Activities	ENABLING Activities	Project being implemented
Africa : CHAD	Second National Communication to the UNFCCC	Climate Change - Mitigation	CC: Enabling Activities	ENABLING Activities	Project being implemented
Africa : CAMEROON, REPUBLIC OF	Sustainable Forest Management by Communities in the Bamenda Highlands, Cameroon.	Biodiversity	3: Forest ecosystems	IDEA	Project has been completed
Africa : COMOROS	Island biodiversity and participatory conservation in the Federal Islamic Republic of the Comoros	Biodiversity	3: Forest ecosystems	IDEA	Project has been completed

(Contd.)

(Contd. Annexure-II)

Region or Country	*Project Title*	*Primary Focal Area*	*Operational Programme*	*Project Type*	*Project Status*
Africa : COMOROS	Capacity Needs Assessment for the implementation of the National Biodiversity Strategy and support to the Clearing House Mechanism	Biodiversity	BD: Enabling Activities	ENABLING Activities	Project being implemented
Africa : COMOROS	Initial Assistance to the Union of the Comoros for Enabling Activities to Implement the Stockholm Convention on POPs	POPS		ENABLING Activities	Project being implemented
Africa : COMOROS	National Capacity Self-Assessment for Global Environmental Management	Multiple Focal Area	Capacity Enabling Activities	ENABLING Activities	Project being implemented
Africa : COMOROS	Capacity Building for Sustainable Land Management in Comoros	Land Degradation	15: Sustainable Land Management	MEDIUM Size Projects	Project being implemented

(Contd.)

(Contd. Annexure-II)

Region or Country	*Project Title*	*Primary Focal Area*	*Operational Programme*	*Project Type*	*Project Status*
Africa : CAPE VERDE	National biodiversity strategy and action plan and country report to the COP	Biodiversity	BD: Enabling Activities	IDEA	Project has been completed
Africa : CAPE VERDE	Enabling Cape Verde to prepare its first national communication in response to its commitments to the UNFCCC	Climate Change - Mitigation	CC: Enabling Activities	ENABLING Activities	Project being implemented
Africa : CAPE VERDE	Integrated participatory ecosystem management in and around protected areas; Phase I	Biodiversity	1: Arid and semi-arid ecosystems	FULL Size Projects	Project being implemented

(Contd.)

(Contd. Annexure-II)

Region or Country	*Project Title*	*Primary Focal Area*	*Operational Programme*	*Project Type*	*Project Status*
Africa : CAPE VERDE	National Capacity Self-Assessment for Global Environment Management (NCSA)	Multiple Focal Area	Capacity Enabling Activities	ENABLING Activities	Project being implemented
Africa : CAPE VERDE	National Adaptation Programme of Action (LDCs)	Climate Change - LDCF	CC: Enabling Activities	IDEA	Project being developed
Africa : CAPE VERDE	Second Communication National to the UNFCCC	Climate Change - Mitigation	CC: Enabling Activities	ENABLING Activities	Project being implemented
Africa : CAPE VERDE	Consolidation of Cape Verde's protected area system through a community-based, participatory approach	Biodiversity	1: Arid and semi-arid ecosystems	IDEA	Project being developed

(Contd.)

(Contd. Annexure-II)

Region or Country	*Project Title*	*Primary Focal Area*	*Operational Programme*	*Project Type*	*Project Status*
Africa : EQUATORIAL GUINEA	National Biodiversity strategy and Action Plan and Country Report to the COP	Biodiversity	BD: Enabling Activities	IDEA	Project has been completed
Africa : EQUATORIAL GUINEA	Strengthening Equatorial Guinea Protected Areas System	Biodiversity		IDEA	Project being developed
Africa : ERITREA	Wind Energy Applications in Eritrea	Climate Change - Mitigation	6: Renewables	FULL Size Projects	Project being implemented
Africa : ERITREA	Conservation management of Eritrea's coastal, marine and island biodiversity	Biodiversity	2: Coastal, marine and freshwater ecosystems	FULL Size Projects	Project being implemented

(Contd.)

(Contd. Annexure-II)

Region or Country	*Project Title*	*Primary Focal Area*	*Operational Programme*	*Project Type*	*Project Status*
Africa : ERITREA	Enabling Eritrea to prepare its first national communication in response to its commitments to the UNFCCC	Climate Change - Mitigation	CC: Enabling Activities	IDEA	Project has been completed
Africa : ERITREA	National Adaptation Plan of Action (LDCs)	Climate Change - Mitigation	CC: Enabling Activities	IDEA	Project has been completed
Africa : ERITREA	Capacity Building for Sustainable Land Management in Eritrea	Land Degradation	CC: Enabling Activities	IDEA	Project being developed
Africa : ETHIOPIA	Eritrea SNC to the UNFCCC	Climate Change - Mitigation		ENABLING Activities	Project being implemented

(Contd.)

(Contd. Annexure-II)

Region or Country	*Project Title*	*Primary Focal Area*	*Operational Programme*	*Project Type*	*Project Status*
Africa : ETHIOPIA	A Dynamic farmer-based approach to the conservation of African Plant Genetic Resources	Biodiversity	3: Forest ecosystems	IDEA	Project has been completed
Africa : ETHIOPIA	National biodiversity strategy and action plan and country report to the COP	Biodiversity	BD: Enabling Activities	IDEA	Project has been completed
Africa : ETHIOPIA	Strengthening the conservation and management of the wildlife protected area system of Ethiopia	Biodiversity	1: Arid and semi-arid ecosystems	PDF B	Project being developed
Africa : ETHIOPIA	Ethiopia: Climate Change Enabling Activity (additional financing for capacity building in priority areas)	Climate Change - Mitigation	CC: Enabling Activities	ENABLING Activities	Project being implemented

(Contd.)

(Contd. Annexure-II)

Region or Country	*Project Title*	*Primary Focal Area*	*Operational Programme*	*Project Type*	*Project Status*
Africa : ETHIOPIA	Expedited Funding for the Preparation of a National Adaptation Programme of Action (NAPA)	Climate Change - LDCF	CC: Enabling Activities	ENABLING Activities	Project being implemented
Africa : ETHIOPIA	Mainstreaming Agro-Biodiversity Conservation into the Farming Systems of Ethiopia	Biodiversity	13: Agro Biodiversity	IDEA	Project being developed
Africa : ETHIOPIA	Coping with Drought and Climate Change	Climate Change - SCCF	17: Climate Change Adaptation	IDEA	Project being developed
Africa : GABON	Conservation of biodiversity through effective management of wildlife trade	Biodiversity	3: Forest ecosystems	IDEA	Project has been completed

(Contd.)

(Contd. Annexure-II)

Region or Country	*Project Title*	*Primary Focal Area*	*Operational Programme*	*Project Type*	*Project Status*
Africa : GABON	National biodiversity strategy and action plan and country report to the COP	Biodiversity	BD: Enabling Activities	ENABLING Activities	Project has been completed
Africa : GABON	Enabling Gabon to fulfill its commitments to the UNFCCC	Climate Change - Mitigation	CC: Enabling Activities	ENABLING Activities	Project being implemented
Africa : GABON Cameroon, Republic of, Congo, Gabon	Conservation of Transboundary Biodiversity in the Minkébé-Odzala-Dja Inter-zone in Gabon, Congo and Cameroon	Biodiversity	3: Forest ecosystems	FULL Size Projects	Project being implemented
Africa : GABON	National Capacity Self-Assessment	Multiple Focal Area	Capacity Enabling Activities	ENABLING Activities	Project being implemented

(Contd.)

(Contd. Annexure-II)

Region or Country	*Project Title*	*Primary Focal Area*	*Operational Programme*	*Project Type*	*Project Status*
Africa : GABON	Second National Communication to the UNFCCC	Climate Change - Mitigation	CC: Enabling Activities	ENABLING Activities	Project being implemented
Africa : GAMBIA	Development of the Gambia's climate change action plan and communications to the UNFCCC	Climate Change - Mitigation	CC: Enabling Activities	ENABLING Activities	Project being implemented
Africa : GUINEA-BISSAU	Guinea Bissau's national biodiversity strategy and action plan and country report to the COP	Biodiversity	BD: Enabling Activities	ENABLING Activities	Project has been completed
Africa : GUINEA-BISSAU	Enabling Guinea-Bissau to prepare its commitments to the UN Framework Convention on Climate Change	Climate Change - Mitigation	CC: Enabling Activities	ENABLING Activities	Project being implemented

(Contd.)

(Contd. Annexure-II)

Region or Country	*Project Title*	*Primary Focal Area*	*Operational Programme*	*Project Type*	*Project Status*
Africa : GUINEA-BISSAU	NCSA Guinea Bissau	Multiple Focal Area	Capacity Enabling Activities	ENABLING Activities	Project being implemented
Africa : GUINEA-BISSAU	Formulation of the National Adaptation Action Plan (NAPA)	Climate Change - LDCF	CC: Enabling Activities	ENABLING Activities	Project being implemented
Africa : GUINEA-BISSAU	Technical, Institutional, Human, and Financial Capacity Building Support for SLM/CCD in Guinea-Bissau	Land Degradation	15: Sustainable Land Management	MEDIUM Size Projects	Project being implemented
Africa : GUINEA-BISSAU	Second national Communication to the UNFCCC	Climate Change - Mitigation	CC: Enabling Activities	ENABLING Activities	Project being implemented

(Contd.)

(Contd. Annexure-II)

Region or Country	*Project Title*	*Primary Focal Area*	*Operational Programme*	*Project Type*	*Project Status*
Africa : GHANA	Renewable Energy-based Electricity for Rural, Social and Economic Development	Climate Change - Mitigation	6: Renewables	FULL Size Projects	Project being implemented
Africa : GHANA Cote d'ivoire, Ghana, Guinea, Liberia, Sierra Leone	Upper Guinea Forest Zone - Participatory Priority Setting and BD Conservation Project	Biodiversity	3: Forest ecosystems	IDEA	Project has been completed
Africa : GHANA Ghana, Kenya, Mali, Zimbabwe	Building capacity in sub-Saharan Africa to respond to the UN Framework Convention on Climate Change	Climate Change - Mitigation	5: Conservation and Efficiency	IDEA	Project has been completed

(Contd.)

(Contd. Annexure-II)

Region or Country	*Project Title*	*Primary Focal Area*	*Operational Programme*	*Project Type*	*Project Status*
Africa : GHANA	Ghana: Preparation of Initial Communication for the Implementation of the UN Framework Convention on Climate Change (UNFCCC)	Climate Change - Mitigation	CC: Enabling Activities	ENABLING Activities	Project being developed
Africa : GHANA Angola, Benin, Cameroon, Republic of, Congo, Congo Democratic Republic, Cote D'ivoire, Equatorial Guinea, Gabon, Liberia, Nigeria, Sao Tome and Principe, Sierra Leone	Combating Living Resource depletion and coastal area degradation in the Guinea Current LME through ecosystem-based regional actions	International Waters	9: Integrated Land and Water	FULL Size Projects	Project being implemented
Africa : GHANA	MS: Biodiversity conservation of Lake Bosumtwe Basin	Biodiversity	3: Forest ecosystems	IDEA	Project has been completed

(Contd.)

(Contd. Annexure-II)

Region or Country	*Project Title*	*Primary Focal Area*	*Operational Programme*	*Project Type*	*Project Status*
Africa : GHANA	National Capacity Self Assessment	Multiple Focal Area	Capacity Enabling Activities	ENABLING Activities	Project being implemented
Africa : GHANA	Sustainable Land Management for Mitigating Land Degradation, Enhancing Agricultural Biodiversity and Reducing Poverty (SLaM) in Ghana.	Land Degradation	15: Sustainable Land Management	MEDIUM Size Projects	Project being implemented
Africa : GHANA	Second National Communication to the UNFCCC	Climate Change - Mitigation		ENABLING Activities	Project being implemented
Africa : GUINEA	National biodiversity strategy and action plan and country report to the COP	Biodiversity	BD: Enabling Activities	IDEA	Project has been completed

(Contd.)

(Contd. Annexure-II)

Region or Country	*Project Title*	*Primary Focal Area*	*Operational Programme*	*Project Type*	*Project Status*
Africa : GUINEA	Enabling Guinea to fullfil its commitments to the UN Framework Convention on Climate Change	Climate Change - Mitigation	CC: Enabling Activities	ENABLING Activities	Project being implemented
Africa : GUINEA	Conservation of the biodiversity of the Nimba Mountains through integrated and participatory management	Biodiversity	4: Mountain ecosystems	FULL Size Projects	Project being implemented
Africa : GUINEA	National Capacity Needs Self-Assessment for Global and National Environment Management in Guinea	Multiple Focal Area	Capacity Enabling Activities	ENABLING Activities	Project being implemented

(Contd.)

(Contd. Annexure-II)

Region or Country	Project Title	Primary Focal Area	Operational Programme	Project Type	Project Status
Africa : GUINEA	Expedited Funding of National Adaptation Programmes of Action (NAPAs) Least Developed Countries	Climate Change - LDCF	CC: Enabling Activities	ENABLING Activities	Project being implemented
Africa : GUINEA	Institutional Capacity Building to Combat Desertification and Land Degradation in the Republic of Guinea	Land Degradation	15: Sustainable Land Management	MEDIUM Size Projects	Project being implemented
Africa : COTE dIVOIRE	Control of Aquatic Weeds to enhance and restore biodiversity	Biodiversity	2: Coastal, marine and freshwater ecosystems	FULL Size Projects	Project being implemented
Africa : COTE dIVOIRE Benin, Cameroon, Republic	Industrial water pollution control in the Gulf of Guinea Large	International	8: Waterbody	IDEA	Project has been completed

(Contd.)

(Contd. Annexure-II)

Region or Country	*Project Title*	*Primary Focal Area*	*Operational Programme*	*Project Type*	*Project Status*
of, Cote d'ivoire, Ghana, Nigeria, Togo	Marine Ecosystem Control of greenhouse gas emissions through energy efficient building technology in West Africa	Waters Climate Change - Mitigation	5: Conservation and Efficiency	FULL Size Projects	Project being implemented
Africa : COTE dIVOIRE	National Capacity Needs Self Assessment for Global and National Environment Management in Côte d'Ivoire	Multiple Focal Area	Capacity Enabling Activities	ENABLING Activities	Project being implemented
Africa : KENYA	Developing Incentives for Community Participation in Forest Conservation Through the Use of Commercial Insects in Kenya	Biodiversity	3: Forest ecosystems	MEDIUM Size Projects	Project being implemented
Africa : KENYA	Institutional support for the protection of East African biodiversity	Biodiversity	3: Forest ecosystems	IDEA	Project has been completed

(Contd.)

(Contd. Annexure-II)

Region or Country	*Project Title*	*Primary Focal Area*	*Operational Programme*	*Project Type*	*Project Status*
Africa : KENYA	Removal of barriers to energy conservation and energy efficiency in small and medium scale enterprises	Climate Change - Mitigation	5: Conservation and Efficiency	IDEA	Project has been completed
Africa : KENYA	LADA (UNEP)			IDEA	Project being developed
Africa : KENYA	Improved Conservation and Governance for Kenya Coastal Forest Protected Area System	Biodiversity	3: Forest ecosystems	MEDIUM Size Projects	Project being implemented
Africa : KENYA	Market Transformation for Efficient Biomass Stoves for Institutions and Small and Medium-Scale Enterprises	Climate Change - Mitigation	5: Conservation and Efficiency	MEDIUM Size Projects	Project being implemented

(Contd.)

(Contd. Annexure-II)

Region or Country	*Project Title*	*Primary Focal Area*	*Operational Programme*	*Project Type*	*Project Status*
Africa : KENYA	Using Farmer Field School Approaches to overcome Land Degradation in Agro-pastoralist Areas of Eastern Kenya	Land Degradation	15: Sustainable Land Management	PDF B	Project being developed
Africa : KENYA	Development and Implementation of a Standards and Labelling Programme in Kenya	Climate Change - Mitigation	5: Conservation and Efficiency	IDEA	Project being developed
Africa : KENYA	Coping with Drought and Climate Change	Climate Change - SCCF	17: Climate Change Adaptation	IDEA	Project being developed

(Contd.)

(Contd. Annexure-II)

Region or Country	*Project Title*	*Primary Focal Area*	*Operational Programme*	*Project Type*	*Project Status*
Africa : KENYA	Strengthening the Protected Area Network of the Eastern Montane Forest Hotspots of Kenya	Biodiversity		IDEA	Project being developed
Africa : LESOTHO	Conserving Mountain Biodiversity in Southern Lesotho	Biodiversity	4: Mountain ecosystems	IDEA	Project has been completed
Africa : LESOTHO	National Biodiversity Strategy and Action Plan (BSAP) and Country Report to the Conference of the Parties (COP)	Biodiversity	BD: Enabling Activities	IDEA	Project has been completed
Africa : LESOTHO	Identifying and Overcoming Barriers to Widesperad Adoption of Renewable Energy-Based Rural Electrification in Lesotho	Climate Change - Mitigation	6: Renewables	FULL Size Projects	Project being implemented
Africa : LESOTHO	National Capacity Self Assessment	Multiple Focal Area	Capacity Enabling Activities	ENABLING Activities	Project being implemented

(Contd.)

(Contd. Annexure-II)

Region or Country	*Project Title*	*Primary Focal Area*	*Operational Programme*	*Project Type*	*Project Status*
Africa : LESOTHO	Rehabilitation of degraded lands of Lesotho; Community-Based approach to sustainable land management	Land Degradation	15: Sustainable Land Management	IDEA	Project being developed
Africa : LIBERIA	Liberia's National Biodiversity Strategy and Action Plan and Country Report to the COP	Biodiversity	BD: Enabling Activities	ENABLING Activities	Project being implemented
Africa : LIBERIA	Mainstreaming and Capacity Building for Sustainable Land Management	Land Degradation	15: Sustainable Land Management	MEDIUM Size Projects	Project being implemented
Africa : MADAGASCAR	Madagascar environment program support	Biodiversity	3: Forest ecosystems	IDEA	Project has been completed
Africa : MADAGASCAR	Enabling Madagascar to fullfill its commitments to the UNFCCC - ENABLING	Climate Change - Mitigation	CC: Enabling Activities	ENABLING Activities	Project being implemented
Africa : MADAGASCAR	Participatory Community-based Conservation in the Anjozorobe Forest Corridor	Biodiversity	4: Mountain ecosystems	MEDIUM Size Projects	Project being implemented

(Contd.)

(Contd. Annexure-II)

Region or Country	*Project Title*	*Primary Focal Area*	*Operational Programme*	*Project Type*	*Project Status*
Africa : MADAGASCAR	National Capacity Self Assessment for Global Environmental Management (NCSA).	Multiple Focal Area	Capacity Enabling Activities	PDF A	Project being developed
Africa : MADAGASCAR	Expedited Funding of National Adaptation Programmes of Action (NAPA) by Least Developed Countries	Climate Change - Mitigation	CC: Enabling Activities	IDEA	Project being developed
Africa : MADAGASCAR	Madagascar: Third Environment Programme	Biodiversity	3: Forest ecosystems	FULL Size Projects	Project being implemented
Africa : MADAGASCAR	Madagascar: Network of Managed Resource Protected Areas	Biodiversity		IDEA	Project being developed

(Contd.)

(Contd. Annexure-II)

Region or Country	*Project Title*	*Primary Focal Area*	*Operational Programme*	*Project Type*	*Project Status*
Africa : MAURITIUS	Restoration of highly degraded and threatened native forests in Mauritius	Biodiversity	3: Forest ecosystems	IDEA	Project has been completed
Africa : MAURITIUS	Conservation of Marine Protected Areas	Biodiversity	2: Coastal, marine and freshwater ecosystems	MEDIUM Size Projects	Project being implemented
Africa : MAURITIUS Comoros, Kenya, Madagascar, Mauritius, Mozambique, Seychelles, South Africa, Tanzania, United Republic of	Agulhas and Somali Current Large Marine Ecosystem (ASCLME) Project	International Waters	9: Integrated Land and Water	FULL Size Projects	Project being implemented
Africa : MAURITIUS	Enabling activities for the Stockholm Convention on Persistent Organic Pollutants (POPs): National Implementation Plan for Mauritius	POPS	14: Persistent Organic Pollutants	ENABLING Activities	Project being implemented

(Contd.)

(Contd. Annexure-II)

Region or Country	*Project Title*	*Primary Focal Area*	*Operational Programme*	*Project Type*	*Project Status*
Africa : MAURITIUS	Removal of Barriers to Energy Efficiency and Energy Conservation in Buildings	Climate Change - Mitigation	5: Conservation and Efficiency	MEDIUM Size Projects	Project being implemented
Africa : MAURITIUS	Capacity Building for Sustainable Land Management in Mauritius	Land Degradation	15: Sustainable Land Management	MEDIUM Size Projects	Project being implemented
Africa : MAURITIUS	Expanding Coverage and Strengthening Management Effectiveness of the Terrestrial	Biodiversity	3: Forest ecosystems	PDF B	Project being developed

(Contd.)

(Contd. Annexure-II)

Region or Country	*Project Title*	*Primary Focal Area*	*Operational Programme*	*Project Type*	*Project Status*
Africa : MAURITIUS	Sustainable Management of POPs	POPS	14: Persistent Organic Pollutants	MEDIUM Size Projects	Project being implemented
Africa : MAURITIUS	Decentralised wind electric power for social and economic development (Alizes Electriques)	Climate Change - Mitigation	6: Renewables	IDEA	Project has been completed
Africa : MAURITIUS	BRIEF NCSA Mauritania	Multiple Focal Area		ENABLING Activities	Project being implemented
Africa : MAURITIUS	ADRAR Solar Initiative and Decentralised Electrification in the Northern Coastline of Mauritania through Hybrid (Wind Diesel) Systems	Climate Change - Mitigation	6: Renewables	PDF B	Project being developed

(Contd.)

(Contd. Annexure-II)

Region or Country	*Project Title*	*Primary Focal Area*	*Operational Programme*	*Project Type*	*Project Status*
Africa : MAURITANIA	Partnership to Mainstream Biodiversity into Oil and Gas Sector Development in Mauritania	Biodiversity	2: Coastal, marine and freshwater ecosystems	IDEA	Project being developed
Africa : MALI	National biodiversity strategy and action plan and country report to the COP	Biodiversity	BD: Enabling Activities	IDEA	Project has been completed
Africa : MALI	Enabling Mali to prepare its first national communication in response to its commitment to the UNFCCC	Climate Change - Mitigation	CC: Enabling Activities	ENABLING Activities	Project being implemented
Africa : MALI	Household Energy and Universal Rural Access	Climate Change - Mitigation	6: Renewables	PDF B	Project being developed

(Contd.)

(Contd. Annexure-II)

Region or Country	*Project Title*	*Primary Focal Area*	*Operational Programme*	*Project Type*	*Project Status*
Africa : MALI Gambia, Guinea, Senegal	Sustainable management of globally significant endemic ruminant livestock of West Africa	Biodiversity	13: Agro Biodiversity	FULL Size Projects	Project being implemented
Africa : MALI	National Capacity Self-Assessment for Global Environmental Management in Mali	Multiple Focal Area	Capacity Enabling Activities	ENABLING Activities	Project being implemented
Africa : MALI	National Adaptation Programme of Action (LDCs)	Climate Change - LDCF	CC: Enabling Activities	IDEA	Project has been completed
Africa : MALI Benin, Burundi, Cameroon, Republic of, Central African	PDF B I : First Regional Micro/mini-Hydropower Capacity Development Project and	Climate Change -	6: Renewables	PDF B	Project being developed

(Contd.)

(Contd. Annexure-II)

Region or Country	*Project Title*	*Primary Focal Area*	*Operational Programme*	*Project Type*	*Project Status*
Africa : MALI	Investment in Rural Electricity Access in Sub-Saharan Africa	Mitigation			
Africa : MALI	Seconde Communication Nationale UNFCCC	Climate Change - Mitigation	CC: Enabling Activities	ENABLING Activities	Project being implemented
Africa : MALAWI	Mali Protected Areas	Biodiversity		IDEA	Project being developed
Africa : MALAWI	SIP: Integrated Management of Cotton and Rice Production Landscapes	Land Degradation		IDEA	Project being developed
Africa : MALAWI	Enabling Malawi to prepare its first communication in response to its commitment to the UNFCCC	Climate Change - Mitigation	CC: Enabling Activities	ENABLING Activities	Project has been completed

(Contd.)

(Contd. Annexure-II)

Region or Country	*Project Title*	*Primary Focal Area*	*Operational Programme*	*Project Type*	*Project Status*
Africa : MALAWI	Barrier Removal to Renewable Energy in Malawi (BARREM)	Climate Change - Mitigation	6: Renewables	FULL Size Projects	Project being implemented
Africa : MALAWI	Combatting siltation in Shire River	Land Degradation		IDEA	Project being developed
Africa : MALAWI	National Capacity Self Assessment	Multiple Focal Area	Capacity Enabling Activities	ENABLING Activities	Project being implemented
Africa : MOZAMBIQUE	National Adaptation Plan of Action	Climate Change - LDCF	CC: Enabling Activities	ENABLING Activities	Project being implemented

(Contd.)

(Contd. Annexure-II)

Region or Country	*Project Title*	*Primary Focal Area*	*Operational Programme*	*Project Type*	*Project Status*
Africa : MOZAMBIQUE	Malawi' Second National Communication to the UNFCCC	Climate Change - Mitigation		ENABLING Activities	Project being implemented
Africa : MOZAMBIQUE	Enabling Mozambique to prepare its first National Communication to the UNFCCC	Climate Change - Mitigation	CC: Enabling Activities	ENABLING Activities	Project being implemented
Africa : MOZAMBIQUE	Expedited Funding of National Adaptation Programmes of Action (NAPA) by Least Developed countries	Climate Change - LDCF	CC: Enabling Activities	ENABLING Activities	Project being implemented
Africa : MOZAMBIQUE	Coping with Drought and Climate Change	Climate Change - SCCF	17: Climate Change Adaptation	FULL Size Projects	Project being implemented

(Contd.)

(Contd. Annexure-II)

Region or Country	*Project Title*	*Primary Focal Area*	*Operational Programme*	*Project Type*	*Project Status*
Africa : MOZAMBIQUE	Strengthening Management Effectiveness of the Terrestrial Protected Area Networkin Mozambique	Biodiversity		IDEA	Project being developed
Africa : NAMIBIA Angola, Namibia, South Africa	Intergrated Management of the Benguela Current Large Marine Ecosystem(BCLME)	International Waters	8: Waterbody	FULL Size Projects	Project being implemented
Africa : NAMIBIA Namibia	Barrier Removal to Namibian Renewable Energy Programme-Phase I	Climate Change - Mitigation	6: Renewables	FULL Size Projects	Project being implemented
Africa : NAMIBIA	Namibia: Enabling Activities for the Preparation of Initial National Communication Related to UNFCCC	Climate Change - Mitigation	CC: Enabling Activities	ENABLING Activities	Project being implemented
Africa : NAMIBIA	National Capacity Self Assessment	Multiple Focal Area	CapacityEnabling Activities	ENABLING Activities	Project being implemented

(Contd.)

(Contd. Annexure-II)

Region or Country	*Project Title*	*Primary Focal Area*	*Operational Programme*	*Project Type*	*Project Status*
Africa : NAMIBIA	Barrier Removal to Namibian Renewable Energy Programme (NAMREP) Phase II	Climate Change - Mitigation	6: Renewables	FULL Size Projects	Project being implemented
Africa : NAMIBIA	Strengthening the Protected Area Network	Biodiversity	1: Arid and semi-arid ecosystems	FULL Size Projects	Project being implemented
Africa : NAMIBIA	Country Pilot Partnership for Integrated Sustainable Land Management	Land Degradation	15: Sustainable Land Management	FULL Size Projects	Project being implemented
Africa : NAMIBIA	Distance Learning and Information Sharing Tool for the Benguela Coastal Areas (DLIST – Benguela)	International Waters	9: Integrated Land and Water	MEDIUM Size Projects	Project being implemented
Africa : NAMIBIA	Development and Implementation of the Strategic Action Programme for the Orange-Senqu River Basin.	International Waters	9: Integrated Land and Water	PDFB	Project being developed
Africa : NAMIBIA	Enabling Activity for the Preparation of Namibia's Second National Communication to the UNFCCC	Climate Change - Mitigation	CC: Enabling Activities	ENABLING Activities	Project being implemented

(Contd.)

(Contd. Annexure-II)

Region or Country	*Project Title*	*Primary Focal Area*	*Operational Programme*	*Project Type*	*Project Status*
Africa : NAMIBIA	Adapting to Climate Change through the Improvement of Traditional Crops and Livestock farming (CCA)	Climate Change - SPA		MEDIUM Size Projects	Project being implemented
Africa : NAMIBIA	NAM-CPP Enhancing Institutional and Human Resource Capacity through Local Level Coordination of Integrated Rangeland Management and Support (CALLC)	Land Degradation	15: Sustainable Land Management	MEDIUM Size Projects	Project being implemented
Africa : NAMIBIA	Sustainable Land Management Support and Adaptive Management (CPP Umbrella) Project	Land Degradation	15: Sustainable Land Management	FULL Size Projects	Project being implemented
Africa : NAMIBIA	Protected Landscapes Conservation Areas Initiative	Biodiversity		IDEA	Project being developed
Africa : NIGER benin, burkina faso, cameroon, republic of, chad, guinea, mali, niger, nigeria	Reversing land and water degradation trends in the Niger River Basin	International Waters	9: Integrated Land and Water	FULL Size Projects	Project being implemented

(Contd.)

(Contd. Annexure-II)

Region or Country	*Project Title*	*Primary Focal Area*	*Operational Programme*	*Project Type*	*Project Status*
Africa : NIGER	Niger's national biodiversity strategy and action plan and country report to the COP	Biodiversity	BD: Enabling Activities	ENABLING Activities	Project has been completed
Africa : NIGER	Enabling Niger to prepare its first national communication in response to its commitment to the UNFCCC	Climate Change - Mitigation	CC: Enabling Activities	ENABLING Activities	Project being implemented
Africa : NIGER	Co-Management of Resources in the Air and Tenere National Nature Reserve and adjacent areas (COGERAT)	Land Degradation	15: Sustainable Land Management	FULL Size Projects	Project being implemented
Africa : NIGER	National Capacity Self Assessment	Multiple Focal Area	Capacity Enabling Activities	IDEA	Project has been completed
Africa : NIGER	Formulation of a National Adaptation Programme of Action to Climate Change (NAPA)	Climate Change - LDCF	CC: Enabling Activities	ENABLING Activities	Project being implemented
Africa : NIGER	Control of Sand Invasion of the Oasis MicroBasins of the Regions of Gouré and Maine (PLECO).	Land Degradation	15: Sustainable Land Management	PDF B	Project being developed

(Contd.)

(Contd. Annexure-II)

Region or Country	*Project Title*	*Primary Focal Area*	*Operational Programme*	*Project Type*	*Project Status*
Africa : NIGER	Second National Communication to the UNFCCC	Climate Change - Mitigation	CC: Enabling Activities	ENABLING Activities	Project being implemented
Africa : NIGER	Niger Fauna Corridors	Biodiversity		IDEA	Project being developed
Africa : NIGER	Nigeria BD Enabling Activities	Biodiversity	BD: Enabling Activities	IDEA	Project has been completed
Africa : NIGER	Enabling Activities for the Implementation of the Convention on Climate Change (UNFCCC)	Climate Change - Mitigation	CC: Enabling Activities	ENABLING Activities	Project being implemented

(Contd.)

(Contd. Annexure-II)

Region or Country	*Project Title*	*Primary Focal Area*	*Operational Programme*	*Project Type*	*Project Status*
Africa : NIGERIA	Self Assessment Exercise_SNC to UNFCCC.	Climate Change - Mitigation	CC: Enabling Activities	PDF A	Project being implemented
Africa : NEW YORK	Support Programme for Biodiversity Enabling Activities	Biodiversity	BD: Short Term	PDF B	Project has been completed
Africa : NEW YORK Burundi, Congo Democratic Republic, Tanzania, United Republic of, Zambia	Partnership Interventions for the Implementation of Strategic Action Programme (SAP) for Lake Tanganyika	International Waters	9: Integrated Land and Water	FULL Size Projects	Project being implemented
Africa : NEW YORK	Africa Land and Water Initiative - Demonstration phase - Institutional strengthening and resource mobilisation (WB, UNDP, UNEP)	Integrated Ecosystem Management	12: Ecosystem Management	MEDIUM Size Projects	Project being implemented

(Contd.)

(Contd. Annexure-II)

Region or Country	*Project Title*	*Primary Focal Area*	*Operational Programme*	*Project Type*	*Project Status*
Africa : NEW YORK Benin, Burkina Faso, Burundi, Chad, Cote d'ivoire, Gabon, Gambia, Ghana, Guinea (french), Mali, Niger, Nigeria, Senegal, Togo	Capacity Building for Improving the Quality of Greenhouse Gas Inventories (West and Franco phone Central Africa)	Climate Change - Mitigation	CC: Enabling Activities	FULL Size Projects	Project being implemented
Africa : NEW YORK	Strategic Investment Programme for in Sub-Saharan Africa (SIP) (Phase 1)	Land Degradation	15: Sustainable Land Management	IDEA	Project being developed

(Contd.)

(Contd. Annexure-II)

Region or Country	*Project Title*	*Primary Focal Area*	*Operational Programme*	*Project Type*	*Project Status*
Africa : CONGO	Enabling Congo to fulfill its commitments to the UNFCCC	Climate Change - Mitigation	CC: Enabling Activities	ENABLING Activities	Project being implemented
Africa : CONGO	National biodiversity strategy and action plan and country report to the COP	Biodiversity	BD: Enabling Activities	IDEA	Project has been completed
Africa : CONGO	National Capacity Self-Assessment	Multiple Focal Area	Capacity Enabling Activities	ENABLING Activities	Project being implemented
Africa : CONGO	Second National Communication to the UNFCCC	Climate Change - Mitigation	CC: Enabling Activities	ENABLING Activities	Project being implemented
Africa : CONGO congo democratic republic	Conservation and sustainable management of the transboundary wetland of Lake Tele/Lake Tumba	Biodiversity		IDEA	Project being developed

(Contd.)

(Contd. Annexure-II)

Region or Country	*Project Title*	*Primary Focal Area*	*Operational Programme*	*Project Type*	*Project Status*
Africa : RWANDA	National biodiversity strategy and action plan and country report to the COP	Biodiversity	BD: Enabling Activities	ENABLING Activities	Project has been completed
Africa : RWANDA	Conservation of the Montane Forest Protected Area System in Rwanda	Biodiversity	4: Mountain ecosystems	FULL Size Projects	Project being implemented
Africa : RWANDA	Improving agricultural extension services for sustainable land use management in Rwanda	Land Degradation	15: Sustainable Land Management	MEDIUM Size Projects	Project being implemented
Africa : SOUTH AFRICA Angola, Botswana, Lesotho, Malawi, Mozambique, Namibia, South Africa, Swaziland, Zambia, Zimbabwe	Capacity building network for southern African Botanical diversity	Biodiversity	3: Forest ecosystems	IDEA	Project has been completed

(Contd.)

(Contd. Annexure-II)

Region or Country	*Project Title*	*Primary Focal Area*	*Operational Programme*	*Project Type*	*Project Status*
Africa : SOUTH AFRICA	Development and Implementation of the National Biodiversity Strategy and Action Plan (BSAP) in South Africa.	Biodiversity	BD: Enabling Activities	ENABLING Activities	Project being implemented
Africa : SOUTH AFRICA	Solar Water Heaters (SWH) for Urban Housing in South Africa	Climate Change - Mitigation	6: Renewables	MEDIUM Size Projects	Project being implemented
Africa : SOUTH AFRICA	Pilot Production and Commercial Dissemination of Solar Cookers in South Africa	Climate Change - Mitigation	6: Renewables	IDEA	Project has been completed

(Contd.)

(Contd. Annexure-II)

Region or Country	Project Title	Primary Focal Area	Operational Programme	Project Type	Project Status
Africa : SOUTH AFRICA	South Africa Wind Energy Programme (SAWEP)	Climate Change - Mitigation	6: Renewables	FULL Size Projects	Project being implemented
Africa : SOUTH AFRICA	Conservation and Sustainable Use of Biodiversity on the South African Wild Coast	Biodiversity	2: Coastal, marine and freshwater ecosystems	FULL Size Projects	Project being implemented
Africa : SOUTH AFRICA	C.A.P.E Agulhas Biodiversity Initiative (ABI)	Biodiversity	1: Arid and semi-arid ecosystems	FULL Size Projects	Project being implemented
Africa : SOUTH AFRICA	C.A.P.E. Biodiversity Conservation and Sustainable Development	Biodiversity		FULL Size Projects	Project being implemented

(Contd.)

(Contd. Annexure-II)

Region or Country	*Project Title*	*Primary Focal Area*	*Operational Programme*	*Project Type*	*Project Status*
Africa : SOUTH AFRICA	Best Environmental Practice in the Hosting of the World Summit on Sustainable Development	Multiple Focal Area	Capacity Enabling Activities	IDEA	Project has been completed
Africa : SOUTH AFRICA	National Grasslands Biodiversity Programme	Biodiversity	1: Arid and semi-arid ecosystems	FULL Size Projects	Project being implemented
Africa : SOUTH AFRICA	National Capacity Self-Assessment (NCSA) for Global Environmental Management	Multiple Focal Area	Capacity Enabling Activities	ENABLING Activities	Project being implemented
Africa : SOUTH AFRICA	Sustainable Public Transport and Sport, A 2010 Opportunity	Climate Change - Mitigation	11: Transport	FULL Size Projects	Project being implemented
Africa : SENEGAL	ADD-ON: National Biodiversity strategy and Action Plan and Country Report to the COP	Biodiversity	BD: Enabling Activities	ENABLING Activities	Project has been completed

(Contd.)

(Contd. Annexure-II)

Region or Country	*Project Title*	*Primary Focal Area*	*Operational Programme*	*Project Type*	*Project Status*
Africa : SENEGAL Mauritania, Senegal	Biological Diversity Conservation through Participatory Rehabilitation of the Degraded Lands of the Arid and Semi-Arid Transboundary Areas of Mauritania and Senegal	Biodiversity	1: Arid & semi-arid ecosystems	FULL Size Projects	Project being implemented
Africa : SENEGAL	ADD-ON: Enabling Senegal to prepare its first National Communication to UNFCCC	Climate Change - Mitigation	CC: Enabling Activities	ENABLING Activities	Project being implemented
Africa : SENEGAL	Integrated Ecosystem Management in Four Representative Landscapes in Senegal, Tranche 1	Integrated Ecosystem Management	12: Ecosystem Management	FULL Size Projects	Project being implemented

(Contd.)

(Contd. Annexure-II)

Region or Country	*Project Title*	*Primary Focal Area*	*Operational Programme*	*Project Type*	*Project Status*
Africa : SENEGAL Guinea, Mali, Mauritania, Senegal	Senegal River Basin Water and Environmental Management Project	International Waters	9: Integrated Land and Water	FULL Size Projects	Project being implemented
Africa : SENEGAL	National Capacity Self-Assessment	Multiple Focal Area	Capacity Enabling Activities	PDF A	Project being developed
Africa : SENEGAL	Integrated Ecosystem Management in Four Representative Landscapes in Senegal, Tranche 2	Integrated Ecosystem Management	15: Sustainable Land Management	FULL Size Projects	Project being implemented
Africa : SENEGAL	Senegal IEM, Tranche 3	Integrated Ecosystem Management	12: Ecosystem Management	IDEA	Project being developed
Africa : SENEGAL	Groundnut Basin Soil Management and Regeneration	Land Degradation	15: Sustainable Land Management	FULL Size Projects	Project being implemented

(Contd.)

(Contd. Annexure-II)

Region or Country	*Project Title*	*Primary Focal Area*	*Operational Programme*	*Project Type*	*Project Status*
Africa : SENEGAL Cape Verde, Gambia, Guinea-bissau, Mauritania	Adaptation to Climate Change	Climate Change - SPA	17: Climate Change Adaptation	FULL Size Projects	Project being implemented
Africa : SEYCHELLES	ADD-ON: Enabling Seychelles to prepare its first national communication in response to its commitments to the UNFCCC	Climate Change - Mitigation	CC: Enabling Activities	ENABLING Activities	Project being implemented
Africa : SEYCHELLES	Mainstreaming Biodiversity Management into Production Sector Activities	Biodiversity	2: Coastal, marine and freshwater ecosystems	FULL Size Projects	Project being implemented
Africa : SEYCHELLES	Seychelles National Capacity Self Assessment	Multiple Focal Area	Capacity Enabling Activities	ENABLING Activities	Project being implemented
Africa : SEYCHELLES	Enabling Activities for the Preparation of Seychelles' Second National Communication to the United Nations Framework Convention on Climate Change (UNFCCC)	Climate Change - Mitigation	CC: Enabling Activities	ENABLING Activities	Project being implemented

(Contd.)

(Contd. Annexure-II)

Region or Country	*Project Title*	*Primary Focal Area*	*Operational Programme*	*Project Type*	*Project Status*
Africa : SEYCHELLES	Mainstreaming Biodiversity Management into Production Sector Activities	Biodiversity	2: Coastal, marine and freshwater ecosystems	FULL Size Projects	Project being implemented
Africa : SEYCHELLES	Seychelles National Capacity Self Assessment	Multiple Focal Area	Capacity Enabling Activities	ENABLING Activities	Project being implemented
Africa : SEYCHELLES	Enabling Activities for the Preparation of Seychelles' Second National Communication to the United Nations Framework	Climate Change - Mitigation	CC: Enabling Activities	ENABLING Activities	Project being implemented
Africa : SEYCHELLES	Capacity Development for Sustainable Land Management in Seychelles	Land Degradation	15: Sustainable Land Management	MEDIUM Size Projects	Project being implemented

(Contd.)

(Contd. Annexure-II)

Region or Country	*Project Title*	*Primary Focal Area*	*Operational Programme*	*Project Type*	*Project Status*
Africa : SEYCHELLES	Mainstreaming Prevention and Control Measures for Invasive Alien Species into Trade, Transport and Travel across the Production Landscape	Biodiversity	3: Forest ecosystems	FULL Size Projects	Project being implemented
Africa : SIERRA LEONE	National biodiversity strategy and action plan and country report to the COP	Biodiversity	BD: Enabling Activities	IDEA	Project has been completed
Africa : SIERRA LEONE	Enabling Sierra Leone to respond to the United Nations Framework Conventionon Climate Change (UNFCCC)	Climate Change - Mitigation	CC: Enabling Activities	ENABLING Activities	Project being implemented

(Contd.)

(Contd. Annexure-II)

Region or Country	*Project Title*	*Primary Focal Area*	*Operational Programme*	*Project Type*	*Project Status*
Africa : SIERRA LEONE	Preparation of a National Programme of Action for Adaptation to Climate Change	Climate Change - LDCF	CC: Enabling Activities	ENABLING Activities	Project being implemented
Africa : SIERRA LEONE	Capacity Building for Sustainable Land Management in Sierra Leone	Land Degradation	15: Sustainable Land Management	MEDIUM Size Projects	Project being implemented
Africa : SIERRA LEONE	Sierra Leone Second National Communication to the UNFCCC	Climate Change - Mitigation	CC: Enabling Activities	ENABLING Activities	Project being implemented
Africa : SAO TOME AND PRINCIPE	Sao Tome CC Enabling Activities	Climate Change - Mitigation	CC: Enabling Activities	IDEA	Project has been completed

(Contd.)

(Contd. Annexure-II)

Region or Country	*Project Title*	*Primary Focal Area*	*Operational Programme*	*Project Type*	*Project Status*
Africa : SAO TOME AND PRINCIPE	National Capacity Self-Assessment (NCSA) for Global Environmental Management	Multiple Focal Area	Capacity Enabling Activities	ENABLING Activities	Project being implemented
Africa : SAO TOME AND PRINCIPE	Sao Tome and Principe: Capacity Development and Mainstreaming of Sustainable Land Management	Land Degradation	15: Sustainable Land Management	PDF A	Project being developed
Africa : SAO TOME AND PRINCIPE	Sao Tome and Principe Second National Communication to UNFCCC	Climate Change - Mitigation	CC: Enabling Activities	ENABLING Activities	Project being implemented
Africa : SWAZILAND	National Biodiversity Strategy and Action Plan and Country Report to the COP.	Biodiversity	BD: Enabling Activities	ENABLING Activities	Project has been completed
Africa : SWAZILAND	Enabling Activities for the Implementation of the United Nations Framework Convention on Climate Change	Climate Change - Mitigation	CC: Enabling Activities	ENABLING Activities	Project being implemented

(Contd.)

(Contd. Annexure-II)

Region or Country	*Project Title*	*Primary Focal Area*	*Operational Programme*	*Project Type*	*Project Status*
Africa : SWAZILAND	National Capacity Self Assessment	Multiple Focal Area	Capacity Enabling Activities	ENABLING Activities	Project being implemented
Africa : SWAZILAND	SNC	Climate Change - Mitigation	CC: Enabling Activities	ENABLING Activities	Project being implemented
Africa : TOGO	Enabling Togo to fulfil its commitments to the UN Framework Convention on Climate Change	Climate Change - Mitigation	CC: Enabling Activities	ENABLING Activities	Project being implemented
Africa : TOGO	BRIEF NCSA Togo	Multiple Focal Area	Capacity Enabling Activities	ENABLING Activities	Project being implemented
Africa : TOGO	Expedited Funding of National Adaptation Programme of Action (NAPAs) by Least Developed Countries	Climate Change - LDCF	CC: Enabling Activities	ENABLING Activities	Project being implemented
Africa : TOGO	Togo: Second National Communication to the UNFCCC	Climate Change - Mitigation	CC: Enabling Activities	ENABLING Activities	Project being implemented

(Contd.)

(Contd. Annexure-II)

Region or Country	*Project Title*	*Primary Focal Area*	*Operational Programme*	*Project Type*	*Project Status*
Africa : UGANDA	Uganda photovoltaic pilot project (PV) for rural electrification	Climate Change - Mitigation	6: Renewables	FULL Size Projects	Project being implemented
Africa : UGANDA	Conservation of Biodiversity in the Albertine Rift Forest Protected Areas of Uganda	Biodiversity	3: Forest ecosystems	FULL Size Projects	Project being implemented
Africa : UGANDA	MS: Community Based Conservation of Wetland Biodiversity in Uganda	Biodiversity	2: Coastal, marine and freshwater ecosystems	PDF A	Project being developed
Africa : UGANDA	Bwindi Impenetrable National Park and Mgahinga Gorilla National Park Conservation	Biodiversity	4: Mountain ecosystems	FULL Size Projects	Project being implemented
Africa : Tanzania, United	Electricity, fuel and fertilizer from municipal and industrial waste in	Climate Change -	6: Renewables	IDEA	Project has been

(Contd.)

(Contd. Annexure-II)

Region or Country	*Project Title*	*Primary Focal Area*	*Operational Programme*	*Project Type*	*Project Status*
Republic of Tanzania	Tanzania: a biogas plant for Africa	Mitigation			
Africa : Tanzania, United Republic of	Reducing Land Degradation on the Highlands of Kilimanjaro Region	Land Degradation	15: Sustainable Land Management	PDF B	Project being developed
Africa : Tanzania, United Republic of	The Conservation and Management of the Forests of the Eastern Arc Mountains of Tanzania	Biodiversity	3: Forest ecosystems	FULL Size Projects	Project being implemented
Africa : Tanzania, United Republic of	The Development and Management of the Selous-Niassa Wildlife Corridor in Tanzania	Biodiversity	3: Forest ecosystems	MEDIUM Size Projects	Project being implemented

(Contd.)

(Contd. Annexure-II)

Region or Country	*Project Title*	*Primary Focal Area*	*Operational Programme*	*Project Type*	*Project Status*
Africa : Tanzania, United Republic of Kenya, Tanzania, United Republic of, Uganda	Reducing biodiversity loss at cross borders sites in East Africa.	Biodiversity	3: Forest ecosystems	IDEA	Project has been completed
Africa : Tanzania, United Republic of	Marine protected areas at Mnazi Bay	Biodiversity	2: Coastal, marine and freshwater ecosystems	IDEA	Project has been completed
Africa : Tanzania, United Republic of Tanzania, United Republic of	Transformation of the Rural Photovoltaics (PV) Market in Tanzania	Climate Change - Mitigation	6: Renewables	FULL Size Projects	Project being implemented

(Contd.)

(Contd. Annexure-II)

Region or Country	*Project Title*	*Primary Focal Area*	*Operational Programme*	*Project Type*	*Project Status*
Africa : Tanzania, United Republic of	Preparation of the Second Report to CBD and Establishment of a National Clearing House Mechanism (CHM)	Biodiversity	BD: Enabling Activities	IDEA	Project has been completed
Africa : Tanzania, United Republic of	National Capacity Self Asseessment for Global Environmental Management in Tanzania	Multiple Focal Area	Capacity Enabling Activities	ENABLING Activities	Project being implemented
Africa : Tanzania, United Republic of	Extending the Coastal Forest Protected Area Subsystem	Biodiversity		IDEA	Project being developed
Africa : Tanzania, United Republic of	Mainstreaming climate change in integrated water resources management in Pangani River Basin Tanzania.	Climate Change - SCCF	9: Integrated Land and Water	MEDIUM Size Projects	Project being implemented

(Contd.)

(Contd. Annexure-II)

Region or Country	*Project Title*	*Primary Focal Area*	*Operational Programme*	*Project Type*	*Project Status*
Africa : CONGO DEMOCRATICREPUBLIC	Rehabilitation of Protected Areas in the Democratic Republic of the Congo	Biodiversity	3: Forest ecosystems	FULL Size Projects	Project being implemented
Africa : CONGO DEMOCRATICREPUBLIC	Preparatory Activities for the elaboration of a monograph on climate change in Zaire	Climate Change - Mitigation	CC: Enabling Activities	ENABLING Activities	Project being implemented
Africa : CONGO DEMOCRATICREPUBLIC	Emergency response to the refugee driven biodiversity crisis in Zaire	Biodiversity	3: Forest ecosystems	IDEA	Project has been completed
Africa : CONGO DEMOCRATICREPUBLIC	Democratic Congo BD Enabling Activities -IDEA	Biodiversity	BD: Enabling Activities	IDEA	Project has been completed

(Contd.)

(Contd. Annexure-II)

Region or Country	*Project Title*	*Primary Focal Area*	*Operational Programme*	*Project Type*	*Project Status*
Africa : CONGO DEMOCRATICREPUBLIC	NCSA DRC	Multiple Focal Area	Capacity Enabling Activities	ENABLING Activities	Project being implemented
Africa : CONGO DEMOCRATICREPUBLIC	National Adaptation Programme of Action	Climate Change - LDCF	CC: Enabling Activities	ENABLING Activities	Project being implemented
Africa : CONGO DEMOCRATICREPUBLIC	Renforcement des capacités légales et institutionnelles dans la lutte contre la dégradation des sols et la déforestation en RDC.	Land Degradation	15: Sustainable Land Management	PDF A	Project being developed
Africa : CONGO DEMOCRATICREPUBLIC	Congo DR - POPs Enabling Activity: Initial Assistance for preparation of a National	POPS	14: Persistent Organic Pollutants	ENABLING Activities	Project being implemented

(Contd.)

(Contd. Annexure-II)

Region or Country	*Project Title*	*Primary Focal Area*	*Operational Programme*	*Project Type*	*Project Status*
Africa : ZAMBIA	National biodiversity strategy and action plan and country report to the COP	Biodiversity	BD: Enabling Activities	IDEA	Project has been completed
Africa : ZAMBIA	Reclassification and Effective Management of the National Protected Areas System	Biodiversity	1: Arid and semi-arid ecosystems	FULL Size Projects	Project being implemented
Africa : ZAMBIA	National Capacity Self-Assessment for Global Environmental Management (NCSA)	Multiple Focal Area	Capacity Enabling Activities	ENABLING Activities	Project being implemented

(Contd.)

(Contd. Annexure-II)

Region or Country	*Project Title*	*Primary Focal Area*	*Operational Programme*	*Project Type*	*Project Status*
Africa : ZAMBIA	Preparation of the National Adaptation Programme of Action (NAPA)	Climate Change - LDCF	Capacity Enabling Activities	ENABLING Activities	Project being implemented
Africa : ZAMBIA	Enabling Activities for the Preparation of Zambia's Second National Communication to the United Nations Framework Convention on Climate Change (UNFCCC)	Climate Change - Mitigation	CC: Enabling Activities	ENABLING Activities	Project being implemented
Africa : ZIMBABWE	Assessing Capacity Building Needs for Biodiversity Management and Development, and consultations leading to preparation of 2nd National Report to CBD	Biodiversity	BD: Enabling Activities	ENABLING Activities	Project being implemented

(Contd.)

(Contd. Annexure-II)

Region or Country	*Project Title*	*Primary Focal Area*	*Operational Programme*	*Project Type*	*Project Status*
Africa : ZIMBABWE	Conservation and sustainable use of traditional medicinal plants in Zimbabwe	Biodiversity	3: Forest ecosystems	MEDIUM Size Projects	Project being implemented
Africa : ZIMBABWE	Photovoltaics for household and community use	Climate Change - Mitigation	6: Renewables	IDEA	Project has been completed
Africa : ZIMBABWE	Coping with Drought and Climate Change	Climate Change - SCCF		MEDIUM Size Projects	Project being implemented

Executive Summary

This study is an attempt to **understand** energy implications of an effective water policy on Africa, which extends far beyond the energy embedded in water. It does so by **evaluating** the relation between water conflicts and energy challenges within African states and accomplishes it after **studying** the African water system, and confirms that "**African hydrology** is the present basis of African Union (AU), given the network of international rivers connects almost all African nations together"[1]. The report therefore believes that it is necessary to look at different ways to address trans- boundary water conflicts such as **hydroelectricity cooperation diplomacy** as a tool to address it, and studied the effects of the Western African Power Pool (WAPP) and Southern African Power Pool (SAPP) model.

Author Kitissou in his paper **Politics of Water** and Cooperation has expressed this importance of interdependence of water and energy challenges in a subtle way; it mentions that "After their independence, African countries crafted political systems that translate, more or less, the western concept of nation-state. By so doing, they **naturalized the "nation" and nationalized the "nature"**.

Whatever portion of water happens to be in the territory is treated as a matter of national sovereignty. Largely, such a system has proven inadequate for resolving internal conflict and such an attitude may have lead to water-based conflict".

This study makes an attempt to understand what it really means by naturalization of the nation and nationalization of nature for Africa. It does so by looking at repetition of trans-boundary water conflicts within a state and its **impact on energy security** and economic development. These circumstances, the paper infers might have been responsible for massive deforestation, internal migration, soil erosion and disruption of growth. It is reinforced by the fact that an estimated of no more than 20[2] per cent, and in some countries as little as 5 per cent, of the population in Africa (excluding South Africa and Egypt) have direct access to electricity, 46%[3] do not have safe drinking water, and about few hundred thousands have migrated during a drought, and still use traditional biomass as a means of energy option.

Most of the countries in Africa face chronic shortage of water due to harsh climatic conditions. It relies heavily on seasonal rainfall leading to little or costly generation of electricity. Most countries don't have any domestic resource of fossil fuel for domestic use and are paralyzed by power loss due to bad transmission lines and power theft. This has lead to severe environmental problems such as deforestation, soil erosion, and runoff due to overuse of traditional biomass as means of the only cheap domestic energy option in the continent.

This report has made efforts to understand and quantify the relationship between water and energy, by looking at the clear link between water conflicts and energy challenges within the African states and identifies that if a nation has 1) higher observed water conflicts along with higher energy challenge, it is an indicator of a recovering economy unable to be resilient to manage conflicts and be able to achieve development goals milestone. The paper suggests that it is inappropriate for the international community to expect these failed economies to participate in this conversation, given their first priorities would fall under peacekeeping category.

It also observed that higher water conflicts states have higher illiteracy rate such as Angola, Chad, Democratic Republic of Congo, Rwanda and Sudan; 2) lower water conflicts and higher energy challenges are mostly countries such as Burkina Faso, Namibia, other landlocked economies; 3) higher water conflicts and lower energy challenges such as Côte d'Ivoire, Botswana; and 4) lower water conflicts and lower energy challenges are economies that have stable water and energy resources such as Ghana, and Nigeria. The study defines water conflicts as challenges driven by internal, local or trans-boundary pressures.

International community such as World Bank energy team suggests that scaling up access to energy in Africa means new ways of doing business. UNDP estimates the cost of institutional capacity development for energy access, and external support for immediate capacity development needs could cost, at a minimum, in the order of roughly US$ 20 million per country, i.e. ~ US$ 1 billion for all LDCs. The paper studied barriers towards achieving this objective.

It concluded that the delivery of electricity or energy services will not automatically lead to economic development in Africa. This implies that some additional intervention or efforts are needed in areas such as governance, enabling environment, integrated policies, efficient management, and effective regulatory framework.

The report also looked into threats to energy production as an outcome of variable water supply. This, the report infers might be due to **failed energy reforms** in Africa. **Under capitalization** in infrastructure development sector combined with civil wars in some nations, has lead to **overutilization** of wood fuel as the most reliable energy option. Within that context study explores the energy implications on water use and importance of "**integrated water resource management**" as a policy tool.

References

1. Dr. Kitissou : Politics of Water and Cooperation
2. http://www.un.org/ecosocdev/geninfo/afrec/vol18no4/184electric.htm
3. http://en.wikipedia.org/wiki/Drinking_water

Index